D1015579

The Jaguar's Shadow

The Jaguar's Shadow

Searching for a Mythic Cat

Richard Mahler

Yale University Press *New Haven & London*

Frontispiece: The eyes of the jaguar. Photograph © Carol Farneti Foster.

Set in Monotype Bulmer by Duke & Company, Devon, Pennsylvania.
Printed in the United States of America.

Library of Congress Cataloging-in-Publication Data
Mahler, Richard.
The jaguar's shadow : searching for a mythic cat / Richard Mahler.
p.   cm.
Includes bibliographical references and index.
ISBN 978-0-300-12225-1 (cloth : alk. paper) 1. Jaguar. I. Title.
QL737.C23M245 2009
599.75'5—dc22   2008051628

A catalogue record for this book is available from the British Library.

10 9 8 7 6 5 4 3 2 1

For Susan Seymour Adams, Bob Enfield, and Don Mahler

There are certain things in Nature in which beauty and utility,
artistic and technical perfection, combine in some incomprehensible way.
—Konrad Lorenz

Contents

Illustrations

Preface

I am not a scientist. This is not a science book, even though it touches on natural history and summarizes research. Those seeking detailed information about the biology and ecology of jaguars are better off looking elsewhere. Neither am I an archaeologist or anthropologist, so the same admonition applies for these domains.

I am a layman journalist who writes for a general audience. My goal is to bring the story of a little known and widely misunderstood animal to the public at large. Jaguars have had a long, complex association with humans, and this relationship deserves to be explored in an accessible way before the big spotted cats disappear. The threat of extinction in the wild for this species is very real, and jaguars could vanish from the forest within our lifetimes. Many large carnivores, in fact, face a problematic future. Tigers, for example, were reduced by 2008 to a mere 7 percent of their original range. Let us pause to consider what is at stake for large felines, and how we may change our course while there is still time to avoid a dire outcome.

The chain that links the chapters of this book is my personal quest to encounter a jaguar in the wild. Was I driven by ego and sentimentality? You be the judge. Did I finally see one? You'll have to keep reading in order to find out.

The source material I used ranged from my own direct experience and interviews—some random and others planned—to a review of information available via tapes, DVDs, books, radio, podcasts, blogs, newspapers, magazines, wire services, scientific papers, university theses, lectures, books, photographs, and the Internet. When relying on so many diverse media, one has to be careful in evaluating what may or may not be empirically true, out of date, sheer conjecture, biased opinion, or just plain wrong. I've been cautious in making such evaluations and am hopeful that my language reflects this.

I have tried to be accurate in the direct quotations I attribute to those with whom I spoke. I sometimes used a tape recorder, more often a notebook, and occasionally e-mail in conducting interviews. Some of my exchanges were via telephone, others were in person, many employed computers, and a few comprised old-fashioned "snail mail" letters. On some occasions I had to rely on my memory to reconstruct what was said. Human frailty being what it is, not all of my quotations are rendered precisely as spoken, but I have done my best to preserve the spirit and content of what was said and to present it within the proper context. Where I had serious doubts, I attempted to double-check my quotations with original sources. In a couple of instances, I have changed a person's name to protect him or her from possible recrimination.

The chronology of my quest may be confusing. I looked for jaguars over a ten-year period, with most activity between 2004 and 2006. I have deleted most references to specific dates and changed some sequencing for the sake of clarity. Because I needed to earn a living—this effort was not subsidized—I could devote only modest amounts of time at irregular intervals to my travels. I regret that I could not mount a more elaborate campaign to find a jaguar, but one makes the best of what circumstances offer.

Finally, I will state a few relevant biases up front. From the start, I believed that the jaguar has intrinsic value and deserves a reasonable amount of habitat protection and land management. While I appreciate the pressures wrought by increasing population, expanding resource extraction, growing tourism, and the demands of agriculture, I also believe a balance among all interested parties (including indigenous animals) can be found. Part of my eagerness to write about jaguars stems from a desire to describe the complexity and hard work involved in executing successful strategies both to protect the animal and to meet the fundamental needs of people who live among them.

Jaguars are reluctant celebrities and are perfectly content to remain in the shadows that have served them. The fact that they exist at all today is a tribute to their remarkable intelligence and adaptability. May *Homo sapiens* be as fortunate as our species faces habitat challenges in years ahead.

The Jaguar's Shadow

CHAPTER ONE

"God Almighty, That's a Jaguar!"

SPRAWLED SPREAD-EAGLED on the ground, I am held captive by the citrine eyes of a cat that outweighs me by twenty pounds, thrives on raw flesh, and could—if so inclined—crack my cranium like an eggshell. This lithe carnivore is crouched less than a yard from my face, close enough for me to feel the damp breeze of his exhalation. My nose flares to receive a pungent odor that is decidedly feline: equal parts well-licked fur, rich body oils, and muscle-braided flesh. My peripheral vision registers a restless tail, as twitchy as an angry serpent. I admire the burnished gold of a satin coat, splotched with dark squiggles encircling flecks of coal. Daubs of cream—streaked with black coffee—adorn throat, chin, toes, and belly. I see paws as wide as oven mitts, canines the length of my index finger, and a boxy skull as formidable-looking as an infantry helmet.

The jaguar's scalloped ears are stenciled elegantly with ocher and charcoal. They swivel in my direction and dispassionate eyes lock onto mine: lids widen, apertures open. Round pupils fix on jittery hands. Do they perceive the minuscule vibrations wrought by a city-dweller's racing heart? I remain prone, naïvely clasping camera to chest like a soldier's shield. Just like my Canon, the chain-link fence that encloses the jaguar seems inadequate protection. I am convinced that anything could happen, including a breach of the thin barrier separating man and beast. I ride waves of adrenaline, primal fear mingling with awe. I am frozen in place, trapped between competing impulses to fight or to flee. The shutter clicks, and a shiver ices my spine.

But now the cat's unwavering stare softens as his interest fades. He seems to have accepted me as merely another in the daily parade of anonymous spectators, neither friend nor foe. A moment later the jaguar stands up, tendons tight beneath luxuriant fur. I get a final once-over before this feline issues a low cough, flicks his ears, and walks away. I ascribe an attitude of nonchalance to the animal as he glides behind a tree.

❧❧

Something shifts inside our brains and guts when we face an animal that has the power to kill almost anything at its whim. This was true during my Central America photography assignment despite my prior knowledge that jaguar attacks on people are virtually unknown. Although other great cats on occasion attack humans, the jaguar expert Alan Rabinowitz has declared, "there have been no verified records of man-eating jaguars, and relatively few records of jaguars killing people." The Costa Rican authority Eduardo Carrillo goes a step further: "A jaguar could eat any animal that crosses its path. . . . There are no records, however, that jaguars have ever attacked people in the wild."

Stories about fierce, aggressive jaguars killing people are the stuff of folklore. Nonetheless, a parfait of powerful biochemical compounds had done a number on my limbic system. My higher cortex—which knew I was facing a docile jaguar in a well-run zoo—was bypassed entirely.

But I now knew firsthand the pulse-pounding emotional storm a jaguar could spawn. Honed fangs, crouched posture, intense watchfulness, and razor-sharp claws awaken in us primal, visceral synapses unconsciously accorded an ancient foe. The dormant hunter-gatherer recognizes physical attributes designed for stalking, dispatching, and shredding prey. Even a casual confrontation with a top-of-the-food-chain predator clearly stirs our submerged animal nature. The cat elicits our intense curiosity, clearly, but its capacity to mortally wound us also ignites our impulse to survive. When we see a jaguar as opposed to, say, a dolphin or a zebra, a complicated relationship kicks in. It's one we may go a lifetime without experiencing.

The face-off submerged me in a soup of conflicted feelings. I was simultaneously emboldened, bewitched, and repelled. No big surprise here. Jaguars, like other cats, are alluring animals that we modern humans tend to romanticize. (Perhaps we should blame Walt Disney.) Worse yet, we tend to sentimentalize all felines in an anthropomorphic way, attributing peoplelike emotions or motives to their behavior. But jaguars are not like us. I would spend the next five years learning how very different they are.

My story begins in a brutal, unfriendly landscape for which I have always felt a curious and tender affection. The arid regions of the southwestern United States are inhospitable to most fair-weather creatures, including people. Between southeast Arizona's comparatively moist Chiricahua and Santa Rita mountains—among the region's biggest "sky islands"—lie some of the harshest corners of the Lower Forty-eight. The Chihuahuan and Sonoran deserts extend fingers of sand and stone north from the adjacent Mexican states that lend these badlands their names. The heat-baked expanse is home to plants and animals adapted ingeniously to a unique environment. Life, implausible as we generally know it, manages to flourish here.

As an example, one of my favorite denizens is Couch's spadefoot toad, an oddball amphibian that buries itself with shovel-like hind legs beneath gravel-rich soils, surfacing to feed and frolic only after it feels the low-frequency drumming of steady rain above its head. It takes several days of precipitation to convince the palm-size critters that the desert is wet enough for browsing and carousing. Somehow, they find each other. A few weeks later, when the toads' tadpoles are old enough, they climb from shallow pools and dig into their own deep burrows. All surface moisture soon disappears, replaced by white-hot days and blue-cold nights. The toads, ensconced safely underground, slip into a kind of suspended animation that may persist for two years or more.

These tortilla-flat valleys and machete-sharp peaks astound me with such improbable miracles. The deserts also beckon with their promise of profound alone-time, otherwise almost unattainable in this age of "24/7" connectivity. I feel soothed by the land's dense mantle of silence and deep solitude. Higher elevations, dotted with evergreen oaks, sturdy mesquites, and feathery acacias, remind me more of outback Greece and Andalusian Spain, particularly after summer's drenching monsoons spawn muted, serpentine greenery. I find these redoubts starkly beautiful, though they can be deadly to the unwary. Washes and ridges of jigsaw-puzzle terrain are threaded with miles of indistinct trails known only to wild animals, furtive smugglers, and lifelong cowboys. On March 7, 1996, one such cowboy, mounted on a sturdy mule named Snowy River, was following a pack of baying hounds hot on a fresh scent.

⋘⋙

"The dogs had headed toward Red Mountain," Warner Glenn recalled a few months after the fact. "I was desperately trying to stay within hailing distance. I could hear [my hounds] climbing up the thick, brushy, steep northerly slope. The last I heard they were going over the top."

It was a calm morning in late winter. Lean, silver-haired Warner, his Stetson hat shading a perpetual squint and sun-carved wrinkles, was leading a hunt through the Peloncillo Mountains. This sawtooth range marks the western boundary of "the Boot Heel," a rectangular wedge of nearly unin-habited land that pushes a forgotten corner of New Mexico into old Mexico's Chihuahua. It is a tiny, lonely territory, acquired by the United States as a bonus to the 1853 Gadsden Purchase, which secured a railroad route farther north. Searing summer heat, unpredictable rainfall, jagged escarpments, and thorny cacti shield stoic creatures eking out a challenging existence.

The sixty-year-old guide and fourth-generation rancher fronted a mounted team following a cadre of carefully trained dogs. A Marlboro man look-alike whose six-foot-six frame towers over any animal he rides, War-ner was escorting client Al Kriedeman on the fourth morning of a ten-day hunt. The party was driving to the base of the Peloncillos each dawn from Warner's headquarters at the nearby Malpai Ranch. The goal was to track, bay, and shoot a trophy mountain lion.

(Mountain lions are referred to by various names throughout their geographic range, including puma, cougar, panther, catamount, painter, American lion, mountain screamer, swamp cat, and plains, gray, or silver lion. All are one and the same species, known to science as *Puma concolor*.)

Warner's daughter, Kelly Kimbro, and wrangler Aaron Prudler completed the team that scoured the rock-strewn slopes of the Peloncillos, which march along the poorly marked New Mexico–Arizona state line before dissolving into Mexico's Sierra Madre Occidental. Over nearly four hours and twice as many miles, the mules struggled to keep up with the pack as it pursued what was assumed to be a large "tom" (male) lion.

"I rode out on top of the rim, and below me were some large bluffs," Warner told an interviewer. "I could hear the welcome sound of the hounds about a half-mile below me, and I could see what I thought was a lion." Then came an unfamiliar snarl. The noise was definitely feline, but sounded like neither a mountain lion nor a bobcat.

"I got Snowy River within 50 yards," Warner wrote in his book about the incident, *Eyes of Fire*. Dismounting, "I walked around some thick trees and brush. Looking out, I said aloud to myself, 'God almighty, that's a jaguar!'"

Although he knew the borderlands as well as anyone, this was Warner's first encounter with what he labeled "the most beautiful creature I had ever seen." Standing in full sun was an animal long presumed to be locally extinct. Its presence hadn't been confirmed in the United States in nearly a decade, and not in New Mexico for much longer. It wasn't supposed to be here—indeed, the Rorschach pattern dappling its buff coat seemed camouflage better suited to tropical forest than desert scrub—but there a jaguar stood.

Warner raced back to his mule, yanked a point-and-shoot camera from a saddlehorn pouch, and began snapping pictures. The angry cat eyed its pursuers warily, eager for a chance to escape. When an opening occurred it sprinted a half-mile down canyon before holing up in a cluster of boulders. Cornered by the hounds a few minutes later, the jaguar—a mature male—lashed out.

Warner, eager to fully document the occasion with his camera, had gotten too close. He jerked away in the moment the jaguar charged.

"Maple and Cheyenne met him head-on as I jumped backward," the hunting guide recalled. "[These dogs] saved me from having my lap full of clawing, biting jaguar. I saw him go around the ledge and jump out of sight. Later, I saw the cat heading south at a long trot."

Figure 1.1. Cornered by hunting hounds on March 7, 1996, this adult male found along the Arizona–New Mexico border is believed to be the first live wild jaguar photographed in the United States. This cat was later reported killed in Mexico by a federal police officer. Photo © Warner Glenn

Within half an hour the animal may have slunk into Mexico through the few strands of barbed wire that marked the border. Tracks were found in the Peloncillos over the next eight months, though the cat itself was not seen again north of the frontier. But it already had made history. For only the third time since the 1930s, a free-ranging jaguar's presence had been confirmed in New Mexico. The hound called Maple nursed a broken leg, and two other dogs suffered minor claw wounds. For his losses, the rancher had seventeen photos, the first known pictures ever taken of a live wild jaguar in the United States. (Numerous photos of dead U.S. jaguars exist, most showing a proud sportsman alongside a carcass.)

As the lion hunters headed home, a still-marveling Warner "silently gave thanks, then wondered how long it would be before [a jaguar] returned" to the Southwest.

On a sun-bleached day in 1996, a brief newspaper article about the Peloncillo jaguar sighting derailed the smooth trajectory of my comfortable life. At the time these big cats seemed as far away as the Belize Zoo, where I'd first confronted them face-to-face, clutching my camera and staring into their intimidating faces. The animals thrived, I always had assumed, only in warm, moist places like the Amazon jungle and rain forest parks of Central America. But as a journalist, I knew that the most common assumptions are often the most incorrect. I also had learned from covering news events that life can change in an instant. From one moment to the next, an object or incident that previously held only cursory interest can derail a career— or spark a burning passion. So it was in my case, when the newspaper's inconspicuous Associated Press story lit a fire that ultimately incinerated my bank account, tested my sanity, and jeopardized my health in places so remote that should I have died, no trace of my remains probably would have been found for months—if ever.

I embarked on a quest that proved more enduring than fantasy, stronger than ego, and impossible to grasp in all its dimensions. I cannot say exactly when or how I reached a point of no return, nor does that seem important in retrospect. I simply was moved to take a journey like no other. A switch was flipped. I needed to act.

Such obsession is as universal as it is irrational. Ambrose Bierce defined the obsessed man in his *Devil's Dictionary* as a person who is "vexed by an evil spirit" that is always "walking in his shadow." And while that force may not be uniformly malevolent, it is predictably compelling and invariably unfathomable. Who knows precisely what pushes someone to row alone across an ocean in a tiny boat, to climb a continent's highest peak, or to descend into the maw of an unexplored cave? Ordinary citizens take on challenges like these all the time. Their decisions often appear foolhardy to others, but unavoidable to those making them. "Why not?" asks the adventurer as he or she treks off to peer inside the bubbling crater of a quivering volcano. "I may never get another chance!"

High-risk behaviors always raise unanswerable questions. Who really knows why so many have walked, jogged, bicycled, ballooned, wheelchaired, and even driven lawnmowers across all or large parts of the United States? What prompts someone to scale a sheer cliff clinging only by fingertips, to soar on a hang glider buffeted by fickle winds, or to surf a bone-crushing winter breaker of Oahu's north shore? The impulse to face death-defying adversity may be supremely illogical, but many are changed irrevocably by the experience. So when I decided in my sedentary midlife—seven years after a different sort of adventure had sent me alone into an alpine winter— that I wanted to search for a wild jaguar, my goal seemed no less plausible than many pursued by others.

❧❧

Several days after Warner Glenn's jaguar sighting, the *Santa Fe New Mexican* published a color photo of the cat on the front of an inside section. The article outlined what had happened, enlivened by the rancher's pithy quotes and phenomenal picture. The image was of a cornered and slightly crouched jaguar, eyes wide and ears taut. Clad in a gold coat splashed with ebony, the exotic creature looked out of place among piñon pines and prickly pear cacti.

My first thoughts were no doubt like those of many readers: "What are *jaguars* doing in New Mexico? Don't they live in the tropics? Did this one escape from a zoo or private reserve?"

Such questions betrayed my ignorance of jaguar basics. I did not know, for starters, that these are the New World's largest felids (a word derived from *Felidae,* the classification biologists use for members of the cat family). Up to twice the size of its distant cousin, the leaner and more streamlined mountain lion, a jaguar is noted for its large head, stocky frame, short legs, vise-grip jaws, and oversize paws. One zoologist has said the animal is built like a cross between a Sherman tank and a fire hydrant. But its distinctive multihued pelage is a work of art that recalls the abstract designs of fur coats adorning leopards, tigers, and cheetahs. Toward the end of the *New Mexican* story I was surprised to learn that this big cat somehow had persisted—albeit in small numbers—in the southwestern United States for tens of thousands of years, adapting itself successfully to hunt scarce prey in a parched landscape.

I read on, checking scattered references on my bookshelf and computer. The key facts were surprising. An encyclopedia advised that jaguars seldom wear the all-black coats generally assigned to them by Hollywood. Only an estimated 6 percent have "dark phase" pelt coloring, caused by a protein-related gene mutation that affects the cat's hair color. The coats of such jaguars still have the rosettes and spots characteristic of their species, but the jigsaw shapes can only be seen at certain angles of light. Some scientists believe this melanistic mutation actually may be of benefit because it offers excellent camouflage at night.

I learned why use of the term *black panther* is misguided, as no such cat seems to exist. The term *panther* as applied in the United States and Canada is a common epithet for mountain lion—*Puma concolor*—which many scientists believe does not occur in a melanistic phase. (In England, its former colonies, and a few areas in North America, *panther* may refer to a leopard or a jaguar.) The much smaller bobcat, however, does occasionally occur in a black form, and in poor light it can be confused with bigger felids.

Ordinarily the jaguar's glossy, short coat varies from pale yellow to rusty red, adorned with the inky-black broken circles called rosettes. These, in turn, usually enclose irregular polka dots. Along the jaguar's back and chest these patterns may merge into solid dark lines. The cat's abdomen is cream or buff etched with coal-colored markings, its face an amalgam of ochre, ivory, jet, and gold.

"In color there is a very wide range of diversity" among jaguars, according to Dodd Mead's *New International Encyclopedia,* "from a background of dirty white or yellowish to almost black." No two jaguars are matched exactly in coloring or insignia. The configurations emblazoned on the animals' fur are as unique as fingerprints. What's more, markings on each side of a jaguar's body are distinct from each other.

As I absorbed this knowledge, I wondered how such hefty, idiosyncratically marked animals could live in even the most remote deserts of a nation of more than 300 million people without being noticed. Or perhaps the 1996 sighting had been of a straying animal, like a bird blown off course during its seasonal migration. Maybe this was the first, last, and only wild jaguar to set foot in the United States during my adulthood.

As I read about Warner Glenn's sighting, my feelings shifted in short order from astonishment to envy. With no effort to conceal my arrogance, I asked silently: "How dare this cowpoke enjoy such good fortune?" Never mind that my skills as a tracker were almost nil and I had no firsthand experience in the borderland backcountry. I wanted to trade places with the man in the chaps, spurs, and Stetson. As a practical matter, I knew that he deserved this honor, and over time Warner gained my admiration. But in the spring of 1996, my single-minded interest in jaguars was anything but practical.

I took no comfort in knowing that many hunters, particularly those who tracked mountain lions professionally, would not have spared the startled predator's life. Ranch-country animosity toward large carnivores is real, and Warner kept his discovery a secret initially for this very reason. The rancher disclosed later that he received calls from angry neighbors who were upset that he had not simply shot the exotic visitor and kept his mouth closed. Some Southwest cattlemen worried that even a single free-ranging jaguar, presumably always hungry for fresh meat, posed a threat to their livestock, particularly calves.

"Shoot, shovel, and shut up." So goes the admonition among westerners about how to handle unwanted wildlife protected by the government. But Warner refused to go along. "How could I kill something so gorgeous?" he asked rhetorically. This attitude melted my initial skepticism. Here was a big-game hunter with not only a conscience but a deep appreciation for nature's serendipitous miracles. He knew that large, well-managed ranches, along with designated protected areas, offered the best long-term hope for the survival of such a species. Unfortunately, not everyone felt this way. The jaguar Warner photographed was eventually reported killed—thirty miles south of the border—by one of Mexico's *federales,* the nation's rural police officers.

I tucked the news clipping into a manila folder marked JAGUAR. The article joined scores of files I maintained for my research as an independent journalist. My specialties at the time included natural history, New Mexico, and indigenous cultures of the Western Hemisphere. I knew that jaguars fit into all these categories, but I had no spare time in which to follow tangents, immersed as I was in other projects.

The life of a freelancer may seem exciting to outsiders, but the day-to-day reality can be mundane. I earned a modest living through an eclectic array of writing assignments, from newspaper columns to full-length books, radio scripts to grant applications. Like corporate wage earners, self-employed writers sometimes accept work that pays the bills but is not terribly engaging. Such was the case for me in spring 1996. I also was feeling marooned in my personal life. My girlfriend and I were on the verge of splitting up, and a long, cold winter had infused me with wanderlust. I was more than ready for a change in direction.

Despite attempts to distract myself, the image of the New Mexico jaguar kept popping into my head. I could not stop myself from unfolding the article and staring into the eyes of what seemed like a fairy-tale creature. It was as if someone finally had videotaped Scotland's legendary Loch Ness Monster or captured the yeti of the Himalayas. A chimera had been brought to life.

My enthusiasm was boundless. I kept asking friends, "Did you *see* that picture in the paper? No? Well then, let me show you!"

Here was an alpha predator photographed in its prime, stalking the emptiest sector of my home state. Hooked on the thrill of unparalleled discovery, my curiosity pulled me from such prosaic obligations as a newspaper piece about Santa Fe's bed-and-breakfast inns and a magazine feature about New Mexico's official state tree, the piñon.

The Peloncillo jaguar had stirred something in the depths of my imagination. I knew that one way or another I was going to seek out such an animal in its native habitat. I would step inside this dream and make it real. During stolen hours I collected and reviewed fundamental facts about *Panthera onca*—the science world's technical name for the jaguar—while gleaning information through local libraries and the Internet. Here are some fundamentals I accumulated:

Among all felids, only lions and tigers are larger than jaguars, and pound-for-pound the latter may be stronger. Jaguars are lean, muscular, and agile, with almost no stored fat. A well-fed South American specimen may weigh more than 350 pounds, about double the weight of an Arizona or New Mexico cat. An adult jaguar's body measures up to nine feet long, including a slender tail that accounts for one third of its total length. (When held high by a mother jaguar, the tail's black tip is said to help cubs follow mom through thick vegetation.)

The jaguar is closely related to the similar-looking leopard, which is found only in Africa and Asia. Jaguars, once widely distributed through much of the Western Hemisphere, today occur in a broad, discontinuous swath from the southwestern United States to northern Argentina. These highly adaptable cats live in an astonishing array of habitats: swamps, deserts, beaches, pine forests, scrubby woodlands, and grasslands as well as jungles. In fact, a jaguar seems able to thrive nearly any place it can hide, mate, raise young, find water, and kill sufficient prey. In warm climates, speculated the naturalist A. Starker Leopold in *Wildlife of Mexico,* the big cat takes over the ecological function performed in part by coyotes and wolves in temperate zones, keeping the population of deer and smaller mammals in check.

This predator's hunt generally takes place at night or during the crepuscular hours of dawn and dusk. Driven by a physical need to garner a large number of calories every few days, it must kill often and well in order to survive. Every felid is a meat eater, and this shapes its destiny. An example is the jaw structure of *Panthera onca,* which has evolved to be among the strongest of any cat. A large male jaguar is believed to exert up to nine hundred pounds per square inch of tooth pressure. It kills mainly on the

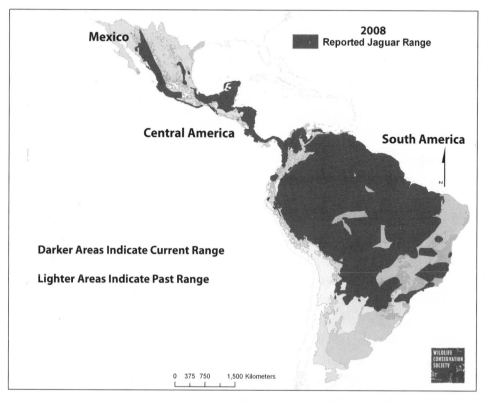

Figure 1.2. The presumed historic range of jaguars extends from arid regions of the southwestern United States to grasslands of northern Argentina. Illustration by Kathy Marieb, © Wildlife Conservation Society

ground from an ambush position, though it also is known to take prey while climbing trees or wading in water.

What's for dinner? The jaguar is known to kill and consume deer, fish, eggs, caimans, capybaras, raccoons, coati, tapirs, turtles, monkeys, birds, snakes, lizards, armadillos, and peccaries. Remains of nearly one hundred species have been detected in the jaguar's feces. Indeed, the cat is believed prone to devour almost any creature that moves, from the swiftest ungulate to the slowest amphibian. Jaguars are opportunistic hunters, even patrolling ocean beaches in search of turtles, and river sandbars where crocodiles sunbathe. In a pinch, they will eat wriggling insects and tiny fledglings. A Mexican hunter who had killed sixty jaguars once told A. Starker Leopold that the contents of no two stomachs had been the same.

Once it identifies a potential victim, the cat tries to get as close as possible while remaining invisible. The Swiss scientist J. R. Rengger's 1830 description of a jaguar stalking a capybara (a large tropical rodent) is particularly vivid: "Serpent-like it winds its way over the ground, pausing for a minute or so to approach it from another direction where there is less risk of being detected. After it has been successful in getting close to its prey, the jaguar pounces on it in one, rarely two, bounds, presses it against the ground, tears out its throat and carries it, still struggling, into a thicket."

Like many cats, jaguars are solitary by nature, with males and females consorting only to mate or during early life as siblings. Adult females are in estrus for six to seventeen days at a time, during which they deposit chemical secretions that signal to mature males their availability for intercourse. Two cubs constitute a standard litter among northern jaguars, while four cubs per birth are common to the south. Newborns are tiny, each weighing two pounds or less. Within weeks their eyes change from china blue to greenish gold. Head and paws grow first; the rest of the body catches up later. Cubs stay with their mothers for eighteen to thirty-six months, dispersing after being weaned and taught hunting and other essential behaviors. The cats typically survive a dozen years in the wild, up to twice that long in captivity. One venerable caged specimen died at thirty-two.

Jaguars are masterful hunters in part because their markings blend easily with vegetation, particularly in darkness or low light. Where it finds plenty of food and a suitable female, a mature adult male may maintain a territorial circuit bounded by ten miles or less, while females patrol a smaller range. Studies suggest that such territories are fluid and vary depending on a host of factors, including prey availability and competition from other predators.

"Cat properties are like ranches," wrote Elizabeth Marshall Thomas in *The Tribe of Tiger,* describing the general behavior of all large felines.

"The space enclosed by the cat's boundaries is actually the grazing land for 'livestock,' whether deer or deer mice, which belongs to the [felid] owner and to no one else, and which the owner does not disturb except to harvest." Young adult males may be great wanderers, dispersing over hundreds of square miles and somehow avoiding detection as they cross highways and pass through yards and pastures.

An animal as extraordinary as this, I wanted to see for myself.

CHAPTER TWO

"It Pays Us Again and Again"

MY LOVE AFFAIR WITH JAGUARS began with a short-lived flirtation in late 1987, when I spent a two-week working vacation in Belize. This tiny Central American nation—the former crown colony of British Honduras, thereby English-speaking—is one of the last strongholds of jaguars. As many as a thousand were thought to remain here in 2009. (The cat also persists in large numbers in much of the Amazon basin and several other heavily forested parts of South America.)

Belize is a haven for jaguars because it is a land where every living creature, including humans, has a good chance of being left alone. "If you feel a need to escape the law, elude creditors, hide assets, or shed the skin of your humdrum life," wrote Bruce Barcott in *The Last Flight of the Scarlet Macaw*, "you could do worse than run away to Belize. Fewer than 300,000 Belizeans spread themselves among the country's river towns and tin-shack villages."

In a region where humans are crowding out wildlife to the point of extirpation—the biologists' term for localized extinction—Belize is an exception to the rule. Relatively abundant here are such fast-disappearing creatures as black howler monkeys, Baird's tapirs, Caribbean manatees, and jabiru storks. About two-thirds of the country is covered by moist broadleaf forest, brackish wetland, and pine-studded savanna. A relative absence of people—the lowest population density in Central America—has been the reality here since around A.D. 900, when the long-dominant Maya civilization fell into a deadly downward spiral. This was a sophisticated culture, whose adherents built impressive limestone cities and mastered complex astronomy. But it imploded and collapsed some five centuries before Columbus first sailed along the Belize coast in 1502.

In 2009 Belize had a population half the size of Albuquerque spread across an area the size of Massachusetts. One-third of its citizens lived in a single community: Belize City. The lack of intrusive development and competition from humans has been a boon for nature's large predators here, although expanding agriculture and increasing population promise to change the situation over time. Immigration from surrounding countries, along with the expansion of timber concessions, oil drilling, and fruit plantations, already has accelerated destruction of native habitat.

I was oblivious to this context when the reggae-blasted public bus in which I was riding trundled across Belize's northern frontier from Chetumal, Mexico. I was en route to the ramshackle, storm-battered sprawl of Belize City. After being ravaged in 1961 by Hurricane Hattie, this coastal town with an alphabet soup of mixed-race residents was displaced as the capital by the made-from-scratch village of Belmopan, fifty miles to the west.

I went to Belize intending to relax, snorkel, and explore Maya ruins, but also to report for a national U.S. magazine on the recent arrival of television. This steamy backwater was dismissed in 1934 by the visiting English writer Aldous Huxley with these caustic remarks: "If the world had any ends, British Honduras would certainly be one of them. It is not on the way from anywhere to anywhere else."

This was one of the last places on Earth to adopt TV. In 1987 a "telly" was still a high-status luxury among Belizeans, and prime-time celebrities familiar to millions around the world remained unknown here. But there would be no turning back. Soon this planetary anomaly would be swept away. My editor felt an account of the transition was necessary.

In an upstairs office devoted primarily to insurance sales, I interviewed an unassuming official wearing Bermuda shorts who was in charge of Belizean television. With a laugh, the bureaucrat cheerfully boasted that local broadcasters shamelessly stole signals from U.S. satellites, "continuing our long, proud history of piracy." Belize—founded by seventeenth-century buccaneers—was so insignificant, he reasoned, that such big-name networks as NBC and HBO would take no punitive legal action.

The insurance agent proceeded to lead me down the street and into a two-story clapboard house where I was introduced to the prime minister himself, a former divinity student named George Price. A devout Catholic who attended Mass each morning, the barefoot but dignified chief of state displayed a picture of the Virgin Mary above his typewriter. Rosary beads lay atop a mahogany desk.

Noticing the notebook clutched in my hand, the gray-haired Creole politician did not mince words: "Are you CIA?"

I smiled at the joke, but the sad-eyed prime minister's expression told me he was as serious as a scorpion's sting.

"If I were CIA, do you think I'd admit it?"

Price shrugged and shot me a wary look. Like many Belizeans, he mistrusted outsiders instinctively, particularly those from the rich countries to the north. They always took more than they gave. Foreign intervention in Central America had led to a string of troubles that at the time of my visit included U.S. military intervention in El Salvador and Nicaragua. During 1954 CIA agents in neighboring Guatemala had overthrown a duly elected president and arranged to have him executed. My effusive denials finally persuaded the prime minister to submit to an hour's worth of questions.

Unfamiliar with the sophistications of modern media, my subject was refreshingly direct and made no effort to soft-pedal his opinions. "By my reckoning, television is a load of steaming crap," Price declared. "I'm so vexed that it has come here. We need refrigerators more than we need boob tubes. I'm proud of my country's high literacy rate—which is almost 100 percent because we love to read and tell stories—but I'm afraid that this may now be in jeopardy."

Without prompting, the prime minister steered the discussion to the economic benefits of conservation, estimating that nature-based tourism was worth more to his country than agriculture and would provide long-term benefits. "We cut a tree and it only pays us once," said Price, wagging a pencil in my direction. "We save that tree and it pays us again and again as foreigners come to enjoy our beautiful rain forests." He mentioned several people I'd planned to interview later in the week, each of whom had a stake in ecotourism. I was surprised that the prime minister knew them all on a first-name basis.

I set down my pen. "Belize is so different from any place I've been," I confessed. "You have a very tight sense of community here; the whole country has the feel of a small town. In fact, I'm surprised you were not aware of who I was before I came through your doorway." The prime minister's sly wink suggested that perhaps he did know—and that I was now privy to a secret revealed to insiders but denied the rest of the world.

"We don't need any spies, or TV news for that matter," he said. "We already have the 'coconut telegraph.' From Ambergris Caye to Punta Gorda, everybody knows what's going on." Suspicions discarded, the avuncular founder of Belize's independence movement stood up, strode from behind his desk, and sent me on my way with a warm bear hug.

A few days later I sipped a piña colada at the Pelican Beach Hotel in Dangriga, a forty-minute plane ride due south. "You simply must go to the Cockscomb," urged the innkeeper Therese Bowman, the twenty-something daughter of an English émigré who had helped establish Belize's prosperous citrus industry. She looked as fresh as an orchid while I wilted in the stifling afternoon heat. The bright-eyed Belizean was describing a newly minted nature reserve called the Cockscomb Basin Wildlife Sanctuary. It was on none of my maps.

"The scenery is spectacular and it's the only place in the world set aside specifically to protect the jaguar," said Therese. A lifelong resident, she spoke in the lilting cadence and modified Oxbridge accent of a British colonial, rendering "jaguar" in three discrete syllables: an elongated *JAG-you-uh* as opposed to the clipped, Americanized *JAG-wahr.* "Prince Philip is an enthusiastic supporter of Cockscomb; he went there a couple of years ago for a dedication ceremony and planted a mahogany tree in front of headquarters."

Therese offered me a second drink in an effort to soothe my frayed nerves after a rough first night in Dangriga. I had come to witness a colorful festival of the local Garifuna, a unique ethnic group descended from African slaves who intermarried with Carib and Arawak tribes. The previous evening a small, unmarked airplane had crashed only one hundred yards from the hotel, scattering plastic packets of freshly harvested marijuana across Dangriga's red-clay landing strip. The uninjured smuggler, apparently flying fast and low from South America, fled on foot into the bushes while cheering locals scooped up contraband. The single-engine Cessna had, with extra gas tanks strapped to its wings, been grossly overweight and run out of fuel.

At Therese's first mention of the jaguar, my brain flashed on an image of the luxury automobile, handcrafted for decades in Coventry, England. I conjured the lunging chrome cat that graces the sports car's sleek, sloping hood.

"I'm not talking about the car," Therese interjected, correctly interpreting my faraway look. "The *cat*."

"Ah, yes," I said. "They're beautiful animals! I'd love to see one." And this was the truth.

On a mist-draped morning I joined a vanload of other tourists on a sightseeing excursion into the Cockscomb Basin, a two-hour drive from Dangriga over wretched dirt roads. I learned en route that the sanctuary was the protected home of two or three dozen jaguars, among the highest known concentrations north of Panama. The 102,000-acre watershed comprised a green curtain of vegetation draping the eastern slope of Belize's highest peaks, sawtooth escarpments used as mariners' landmarks since Spanish flotillas had negotiated outlying coral reefs in the early sixteenth century. Open to the public for only a few months, the park was accessible from the main "highway," itself a miserable muddy track, by more than seven miles of rutted pathway that became nearly impassable muck in wet weather. One of the few straight and level portions of the route had

UNITED STATES

Palm Springs

Childs jaguars, 1996-2008
Glenn jaguars, 1996-2006

Sonora jaguars

MEXICO   Gulf of Mexico

Chichen Itza

La Milpa
Peten   Cockscomb Basin
Bonampak
BELIZE
Antigua   HONDURAS
Moskito Coast
GUATEMALA

EL SALVADOR

NICARAGUA

Primary research conducted   COSTA RICA
January 2004 to January 2009   Panama
                                               Canal
                          Corcovado
                                        PANAMA

Figure 2.1. The author's search for a wild jaguar took him from Arizona and New
Mexico south to below the Panama Canal. Map © Richard Mahler

been bombed intentionally by the Belize Defense Force in a bid—through
the airborne deposit of ordnance craters—to discourage drug-transport
planes from making unauthorized landings. This laid-back country, a stable
democracy with no standing army, was a major transshipment center for
marijuana and cocaine originating in Colombia. (Witness the smashed
Cessna outside my hotel.) No one seemed particularly worried about this,

acknowledging that the drug trade rivaled farming and tourism as a major
revenue producer.

We rolled past bright, crystalline streams frequented by neotropical otters and white-lipped peccaries—both now scarce throughout their
range—and ascended a series of thickly forested hills. The emerald maze
made it difficult to determine compass directions, but I assumed we were
headed west toward Victoria Peak, a 3,675-foot crag that overlooks much
of Belize's famous Caribbean barrier reef.

Our driver, Ramón, jerked his thumb toward a tangle of rusty metal
hanging from a quamwood tree. "That particular plane wasn't carrying
drugs," he advised. "It transported Dr. Alan and the radio equipment he
used to track tigers." Ramón invoked the local name—pronounced *TIE-
gah*—applied to the jaguar, borrowed from a commonly used Spanish
misnomer, *tigre.* He explained that a young wildlife biologist named Alan
Rabinowitz and two others had survived the harrowing 1983 crash with
minor injuries. Rabinowitz had hired the plane in order to follow jaguars
previously harnessed with signal-emitting collars. I later viewed video footage taken inside the cockpit as it crashed into a set of tall trees, clipping a
wing and hurtling the craft onto the forest floor.

Shortly after passing the wrecked plane, our van slid into a jade tunnel of trees so thick with leaves that I could see only a few yards into the
blizzard of vegetation. Here at last was a real jungle, lavishly decorated with
bromeliads, tethered with vines, restless with birds, and smelling concurrently of organic growth and speedy decay.

We pulled into the abandoned lumber camp Rabinowitz had comman-
deered as his headquarters. Quam Bank—named after a commercially valuable species of tree harvested here—once had been home to several Maya families who practiced subsistence farming and hunted wild game. The dainty
brocket deer and a rabbit-sized rodent called agouti were favored targets,
along with turkey-sized birds named curassow and crested guan. The last
two species had nearly been wiped out over the years. The agouti remained
a commonplace culinary delicacy that had earned the nickname "royal rat"
after Queen Elizabeth II was served a grilled one during a state visit.

Near the rustic cabin that once served as Cockscomb's research center lay metal cages used years earlier to capture cats for study. Live pigs had been tied inside the steel rectangles as bait, with mixed results. In order to better his chances of catching jaguars, Rabinowitz eventually hired Bader Hassan, a big-game hunter of regional renown who relied on trained hounds and a handmade device that simulated the cat's distinctive calls. Such "callers" often are fashioned of hollowed gourds or plastic buckets through which a piece of waxed sinew or leather is strung. By rubbing this string in a particular way, sounds akin to those voiced by the jaguar are produced, drawing the curious cat close enough to track with hounds.

Despite many setbacks, including some that were the product of his professional inexperience and admitted arrogance, Rabinowitz's Cockscomb study was deemed a great success. During his two years in the basin, young Rabinowitz, with his assistant Ben Nottingham, compiled more empirical data about the Central American jaguar than had any predecessors. Rabinowitz overcame bouts with tropical diseases and parasites, as well as the death of one research assistant by snakebite. Several radio-collared jaguars also died. His dogged patience and bullheaded strength helped the New Yorker to overcome these hardships and to alert Belizeans about the potentially perilous future of their wild felines. During a fifteen-minute meeting with government leaders he talked them into establishing the world's first jaguar preserve.

"It's never been about just science for me," Rabinowitz, who has the compact strength and lithe movements of the cats he studies, later told an interviewer. "The combination of exploring and being physically challenged and being able to come away with data or make discoveries—to me that is the best of all."

The Cockscomb's now-thick forest was logged selectively into the 1980s, and the Maya farmed and hunted within its boundaries for centuries. Still, I found the mass of foliage at the tallest reach of trees so dense that the ground was dimly lit even at noon. There was barely enough mottled light to allow

Figure 2.2. Alan Rabinowitz, president and CEO of the Panthera Foundation, began his jaguar field research in Belize in 1983 and is now working to save several of the world's largest cats from extinction. Photo © Wildlife Conservation Society

well-adapted ferns, vines, and shrubs to take root in the shallow topsoil. Walking through old-growth areas reminded me of treks I had taken in California redwood groves. Only random patches of blue sky could be seen through thigh-thick branches and foot-wide leaves. I watched dust motes float through shafts of sunlight on a trail eerily silent in noon's sauna-bath heat. My Maya guide, Ignacio Pop, who had worked with Rabinowitz, told me that numerous limestone buildings in the surrounding forest had not been excavated in the millennium since their abandonment. The bones of Pop's ancestors were out there, lying undisturbed.

I was disappointed but not surprised when I failed to see any of the five felid species that Cockscomb protected. Dan Taylor, the Peace Corps volunteer who oversaw the sanctuary at the time, confessed that he had seen "only the disappearing tip of a jaguar's tail" during fourteen months of residence. (Twenty years later, a resident warden told me that about six of the estimated ten thousand annual Cockscomb visitors are lucky enough to glimpse *Panthera onca*.)

My time in the sanctuary was brief. I was at the mercy of our tour guide, who herded us back into his van precisely at the appointed hour. But I had seen enough to know I would return on some future trip to spend several days and nights inside the Cockscomb. I had no inkling how many years it would be before that promise to myself was realized.

After returning from Belize I continued in an offhand fashion to gather information about jaguars. Four years later, in 1990, some of this material appeared in *Belize: A Natural Destination*, a guidebook I cowrote with the conservationist Steele Wotkyns and the photographer Kevin Schafer. Our collaboration promoted the strategy of "ecotourism," which seeks to support local residents and wildlife in part by encouraging foreigners to visit nature reserves. In his preface to our book the permanent secretary of Belize's Ministry of the Environment, Victor González, praised ecotourism's capacity to "produce economic returns to the [region's] people, particularly those who have relied on the forest resources for the livelihood." A year later, a discussion of the jaguar and its role in Maya cosmology worked its way into my companion guide about Guatemala, as well as a series of magazine features I wrote.

The more I learned about the jaguar, the more its preternatural cunning and elusiveness impressed me. Here was one of the few remaining apex predators with sufficient skill, strength, and adaptability to survive almost any challenge, with the notable exceptions of habitat loss and prey disappearance. While its customary homes were tropical forest, tree-studded savanna, and humid wetland, the species had made its home in ecosystems as varied as creosote-dotted deserts, piñon-juniper woodlands, and arboreal forests of spruce, pine, and fir. The jaguar seemed able to thrive in almost any niche, which partially explained its presence in the twenty-first century, long after some great cats—including the Balinese and Javanese tigers—had become extinct. Perhaps jaguars enjoyed advantages these felids had lacked.

"In the struggle for survival," Charles Darwin reminded readers of his *Origin of Species,* "the fittest win out at the expense of their rivals" not simply as a function of brains and brawn, but "because they succeed in adapting themselves best to their environment."

Despite its remarkable ability to adapt, the jaguar was in trouble. This highly evolved loner now competed with human beings—and their impact—nearly everywhere it roamed. Landscape changes and prey loss were ongoing. Despite harsh legal sanctions, hundreds of jaguars were killed by people each year through means as varied as collisions with vehicles and deliberate poisoning. This was particularly true in the Amazon River basin, where livestock graze widely and enforcement of wildlife laws is almost nonexistent. Without large protected spaces containing sufficient game— and wise management of both wildlife habitat and adjacent ranches and farms—it seemed that wild jaguars might not make it into the twenty-second century. In a few places ranchers and farmers had learned to coexist with jaguars, concluding that agricultural use of the landscape was not necessarily incompatible with big cats. Yet effective strategies that took into account the needs of both people and large felids were uncommon. Too often, the solution to conflict was a bullet.

I was shocked to find that not even the experts seemed to know the total population of jaguars. The guesses of scientists varied widely. "There are no reliable figures for the total number of jaguars remaining in the wild," Mel and Fiona Sunquist concluded in *Wild Cats of the World,* published in 2002. "Most countries with jaguar populations have been unable to regulate hunting effectively, and little is known of the population dynamics of wild jaguars."

During the past century, according to reputable surveys, more than half of jaguar habitat has been obliterated or extensively modified. Two-thirds of Central America's original forest cover has been lost and nearly that much in Brazil. Latin America continues to have one of the highest deforestation rates in the world. This in itself is cause of great concern within the conservation community.

"It's not that I want to protect jaguars," the pioneering researcher Alan Rabinowitz once explained. "I'm actually allergic to cats, and every time I capture one my whole face swells up." But in saving an endangered animal that requires a great swath of little-disturbed habitat, "I know I am saving a good ecological system for a variety of species."

I made fourteen trips during the 1980s and 1990s to national parks and nature reserves in Belize, as well as many in neighboring Guatemala, Honduras, and Mexico. But I did not see a single wild jaguar. Though hardly surprising, this outcome disappointed me. Certainly I could write about the cat without encountering one, and I continued to do so.

I knew I could improve the long odds of discovery by hiring a professional like Arizona's Warner Glenn, who hunts with his scent-sensitive hounds an average of 110 days a year, or by tagging along with a wildlife biologist in Latin America. But that felt like cheating. I wanted to see a jaguar without the aid of guides, dogs, horses, mules, callers, lures, traps, sensors, or snares. I rationalized that the accomplishment would be mine only if I achieved it entirely on my own. I knew this would not be easy, but, like the cat itself, I prefer to operate independently. Yet seeing a wild jaguar was only part of the challenge, ancillary and parallel to my primary goal of popularizing the story of *Panthera onca*. There had to be a way to help save the jaguar without jeopardizing the species' survival. The issue was a complicated one, raising ethical questions I needed to mull further.

There had been one close call. I was in northeast Guatemala's Petén jungle, walking alone along an overgrown path at dusk, dinner hour for many

predators. I turned a corner and found myself staring at an equally startled feline. The animal's lean, lithe body came up to my knees, its creamy beige fur
offset by a mosaic of black spots and rosettes. Dark stripes formed a beguil-
ing mask sandwiched between pink nose and translucent ears. Eyes wide,
the ocelot appraised me thoroughly before darting into the underbrush.

Although this chance meeting in Tikal National Park delighted me,
my excitement was tempered somewhat by its occurrence near where a
rescue group was preparing to return a rehabilitated ocelot into the wild.
I had seen the animal pacing around its cage near the hotel's swimming
pool. Apparently the wild ocelot I encountered had been attracted by this
interloper's vocalizing or scent and was closing in for a surreptitious look.
Fiercely territorial, like many cats, a male ocelot may fight to the death in
order to keep a competitor from usurping its home range. The thrill of seeing this cat in its natural surroundings only whetted my appetite.

In January 2001 I met an ardent and gregarious young naturalist named
Mark Pretti at Ramsey Canyon Nature Reserve in the Huachuca range of
southeast Arizona. A tour of the sanctuary had been arranged by my fellow writer Nicky Leach, who planned to interview Pretti for an Arizona
travel guide for which I was also a contributor. Ramsey Canyon encompasses a highly specialized riparian (streamside) habitat not far from the
U.S.-Mexico border. This narrow defile links desert and mountain with
year-round flowing water.

"We are at the northern edge of their range for several species associated with the subtropics," Pretti told us, as we trudged through four inches
of seasonally appropriate snow. "For example, we have lots of coati here. It's
a raccoon relative found more often in lush, moist habitats than arid desert,
yet is pushing farther into Arizona and New Mexico." The sanctuary, Pretti
added, was one of the only places in the country where one could reliably
find emerald and eared trogons. These exotic habitués of Central America
are relatives of the resplendent quetzal, a lovely and reclusive cloud forest
bird known for its exceedingly long tail and iridescent green crest.

We paused before a small pond, partially iced over. Pretti told us this unimpressive body of water, kept filled by spring-fed Ramsey Creek, was home to an exceedingly rare species: the Ramsey Canyon leopard frog. For an animal found only within a four-mile radius of this spot, the humble puddle was the amphibian's last stronghold on Earth.

"What about jaguars?" I interrupted, remembering that Ramsey Canyon and the adjacent Huachucas were within *Panthera onca*'s historic range. "I've read there were a number of sightings in Arizona during the twentieth century." (As many as thirty, according to the Arizona biologist David E. Brown.) "Is there evidence they're still around?"

Pretti shook his head and grinned, as if my question was anticipated. "Nope," he said, "although it's theoretically possible. The last confirmed sighting in our state was in the Baboquívari Mountains, about fifty miles to our west. A lion tracker named Jack Childs, with his wife and friends, took a picture of a jaguar there in August of '96. This was about six months after that rancher found a jaguar in the Peloncillos. The Baboquívaris are more rugged country than this and their western slope is on the Tohono O'odam Indian Nation. I guess if a jaguar is around, that's where you'd find it."

The conversation continued as we walked along the snow-draped trail. Pretti excitedly told Nicky about other animals—the lesser long-nosed bat and beryline hummingbird—found in this narrow canyon but rarely seen anywhere else in the United States. His voice faded away as wheels turned in my head. Something was shifting. A part of me was certain I would take myself to the Baboquívaris and look for a wild jaguar, maybe the same one Jack Childs had seen but perhaps another. I felt my anticipation swell like a pang of hunger. Edgy excitement surfaced as it had when I first read about the Peloncillo sighting made by Warner Glenn. I still wanted to meet—on foot and without guns, gizmos, or hounds—this enduring four-footed symbol of untrammeled New World wilderness.

CHAPTER THREE

"Among All Big Cats, We Know Least About Them"

I DISCOVERED THAT THE AVERAGE person, including me, knows nearly nothing about jaguars. And much of what we *think* we know is wrong. I was surprised that friends and family members had only a vague idea of what I was talking about.

"Which model do you like?" a student, on the verge of receiving a doctorate degree from a prestigious university, inquired.

"They're vicious man-eaters," a cousin declared. "Don't ever turn your back on one because that's when they attack."

"I've seen them at the zoo," a woman at a party enthused. "What beautiful black coats they have!" At the same gathering a retired lawyer insisted that while on safari in East Africa he had watched a jaguar run down a wildebeest.

Several people were convinced the cats were a type of mountain lion: "Cougar? Panther? Jaguar? Aren't they the same thing?"

Patiently, I spelled out some facts. First, a person is much more likely to be killed by a spider, snake, or domestic dog than a jaguar. Second, fewer than 10 percent of jaguars are black. (More on this later.) Third, jaguars are New World cats, found outside the Americas only in captivity. And finally, the jaguar is a distinct species, related only distantly to the felines referred to as cougars and panthers. (More about this later, too.)

When I explained that jaguars occurred historically in the American Southwest and seemed to be drifting back, I was met by skepticism. "Someone is pulling your leg," my next-door neighbor maintained. One of my work colleagues chimed in: "It's a publicity stunt; a hoax! Tame animals are being turned loose in the mountains." Where I saw a marvel, they saw deception. The notion of a wild United States jaguar struck some folks as outlandish, even suspicious. Yet somewhere in Arizona roamed at least one large felid that was as focused as a Zen master, as quiet as a fawn, and as eager to avoid cameras as a movie star in rehab.

❦❦

From the start I was convinced that *Panthera onca,* one of several endangered among the thirty-eight known species of wild cat, was worth saving, if only because it had as much right to exist as did humans. But how would jaguars find protection if 99 percent of the population did not know what they were? Where they lived? How they survived? What made them special? Beyond their charismatic image, these animals were essentially unseen and unknown.

Having reported for years on conservation issues, I knew that people needed to care about a creature before they could be persuaded to help save it. An emotional bond or financial incentive must exist; otherwise, the animal may be deemed insignificant, a competitor for resources, an obstacle to progress, a threat to survival, or a source of fresh meat.

Before I spread the word about jaguars, I had to educate myself further. I felt it was necessary to visit places where the animals were seen in order to speak with scientists who studied these cats and rural residents who lived among them. I had no intention of becoming an authority on jaguars, but I wanted to learn what *real* experts had to say. And I still wanted to observe at least one such cat in the wild.

Accomplishing this would be tricky. Looking for wild jaguars and writing about them posed moral questions. If others followed my example, these animals might become even more endangered. I knew I could not control how my information would be distributed. At worst, my accounts might inspire unscrupulous trophy hunters to track and shoot jaguars. Poachers might use my text as a kind of road map to the cats' strongholds. An increase in nature-based tourism might prompt the cats to withdraw deeper into the wilderness, putting themselves in various kinds of danger.

It was an audacious act to seek out such rare animals merely to satisfy my own curiosity. And wouldn't I be earning money by writing about them? Could that be considered yet another form of exploitation targeting a cat that seemed to want nothing more than to be left alone?

I reflected at length on my underlying motives. My bias was saving jaguars from extinction, an eventuality that already had befallen certain subspecies of tiger. The Cantonese tiger was as good as dead, the Sumatra tiger was critically endangered, and India's tigers were relegated to ever-shrinking islands of habitat. By comparison, the jaguar as a species was in good health.

I implemented damage control. I resolved to withhold certain details about my quest. For instance, I would not reveal specific locations where the most vulnerable cats, including those in the United States, might be found. Even though such coordinates could be divined from other sources, I did not want to be responsible for their circulation. I also would not disclose all that I knew about black-market trafficking of live jaguars or the cats' skins, organs, teeth, claws, and mounted trophies. Hunters and traders still pursue and kill exotic cats, and I would hold back material that might facilitate this. Finally, I was determined to disturb jaguars as little as possible through my physical presence.

Simply *wanting* to see a big cat is insufficient to make it happen, of course. Felines are famous for their smarts and elusiveness. "Jaguars have the biggest brain-to-body mass of all the big cats," Bruce Barcott quoted the Belize researcher Sharon Matola in *The Last Flight of the Scarlet Macaw.* "They're extremely intelligent, which makes them independent, unpredictable, and dangerous. You'll never see a jaguar tamer in the circus. If you do, buy a ticket, because that'll be a one-time show."

Wild cats in general are challenging subjects. For some of the world's smallest, almost nothing is known. The caracal and serval, notoriously bashful African felids, are studied mostly through indirect evidence rather than direct observation. Field researchers learn about wild jaguars in much the same manner.

"Tracking the stealthy, solitary animals," concedes Eduardo Carrillo, a Costa Rican biologist who has encountered more than a dozen in jungle settings, "remains exceedingly difficult." And while there is plenty of general knowledge and considerable data about the jaguar as a species, significant details are missing. It appears, for example, that no one has ever filmed or videotaped au naturel a wild female with her cubs. This fundamental relationship—the foundation upon which the ABCs of "jaguarness" are built—remains something of a mystery. Watching a mother and cub in a zoo can provide clues but not answers, since captive animals may behave differently from their free-range cousins.

Other key questions hang in the air. It is unclear exactly how jaguars, often alone throughout adulthood, find and court their mates. Also, the cats' carefully regulated social structure and precise communication is not well known. Other lines of inquiry beg investigation: How do such alpha predators influence the biodiversity of a given area? Why are jaguars, as compared with most other large cats, such superb, water-loving swimmers? How do jaguars deal with competing carnivores that overlap with them in time, space, and prey? Why does each jaguar have a singular pattern of spots and rosettes, and why is this hodgepodge different on each flank? How many jaguars move between North and South America and how important is such migration in keeping gene pools healthy? If adult males head out from Mexico to far-flung places like the mountains of Arizona, how do they get across a carefully monitored border without being detected? What are they looking for—or fleeing from?

One reason to protect jaguars is simply to follow these and other intriguing lines of research. Knowledge about such highly evolved and specialized creatures may illuminate and improve our own lives. We might learn from the jaguar's finely tuned senses, for example, what is communicable through odors humans cannot smell and sounds we cannot hear. By studying the cat's remarkable night vision and optic system, perhaps we can better understand our own eyesight and the maladies that affect it. Knowing the role such a top predator plays in maintaining a balanced ecosystem could conceivably help us improve management of parks and adjacent agricultural lands. Such investigation is long and arduous, but through perseverance, training, and luck, a growing number of researchers are fitting pieces—a few precious facts at a time—into the jaguar puzzle, often without actually seeing the animals.

❧❧

What's in a name? Jaguar says it all. *Yaguara*—believed to be the root word—comes from a dialect of the Amazon's indigenous Tupi-Guarani people and is translated roughly as "the beast that overcomes its prey in a single bound." It is an apt description, for this is generally how the hemisphere's largest cat feeds itself. In this respect the jaguar is not unlike its pantherine African and Asian cousins: lions, leopards, and tigers. All these felids have strong, thick necks, powerful jaw muscles, and sharp teeth ideally suited for overwhelming large prey with a minimum of struggle. The jawbone itself is short, providing a leverage advantage when sinking canines into a throat or skull. But the jaguar's dentition has evolved in a way that is particularly well suited to snapping vertebrae, crumpling brain cases, crushing windpipes, tearing flesh, and scraping skeletons. Louise H. Emmons wrote in *Cats: Majestic Creatures of the Wild* that the jaguar's "jaws and teeth are so well developed that it is considered the most powerful of all the cats." They are "the only big cats to regularly kill prey by piercing the skull."

The gums and jaw adjacent to the four primary killing teeth of jaguars are threaded with pressure-sensitive nerves that help locate the right spot to execute prey. The cat's digestive tract, like those of other felines, is designed to process meat exclusively. Even its tongue, lined with sharp protuberances, helps remove flesh from cartilage and bone. But unlike African and Asian lions, which stalk, attack, and feed as a group (called a "pride"), the jaguar is a solo hunter that stalks and ambushes on its own.

Any carnivore this large, charismatic, and widely distributed will be assigned many names. The jaguar is no exception, although the mountain lion, with at least sixty-five recorded appellations, probably takes the prize. Beyond the commonly used *tigre* or *tigre pintado* and occasional *leopardo* of Spanish-speaking countries, a jaguar in Portuguese-speaking Brazil is an *onça* (sometimes spelled *onza*). Argentinians call the cat *yaguareté,* while in French Guiana, on South America's northeast shoulder, the moniker is *tig marqué.* Next door, in Surinam, it is *penitigri.* In English-speaking Belize, the jaguar and mountain lion are referred to as *tiger* and *red tiger,* respectively. But no matter what languages they speak, scientists around the world use the Greek-derived *Panthera* (or *Pantera*) *onca,* bestowed in 1758 by the Swedish botanist Carl Linnaeus, the father of taxonomy. The approximate translation of this genus-species assignment is "sharp-clawed hunter of all prey."

A splendid and efficient killing machine, the jaguar is the very embodiment of agility and strength. The cat's finger-length fangs and vise-grip jaws swiftly crush the bones of any animal it attacks, snapping a neck or crushing a skull with ease, causing near-instant death. The jaguar's preferred technique as a bone-crusher differs slightly from the death bites inflicted by mountain lions and some other felids that asphyxiate prey by clamping shut windpipes. A jaguar generally finds it easier to puncture a brain case or sever a spine. Such strong jaws can swiftly break through an armadillo's shell, a turtle's carapace, or a monkey's cranium. "No other great cats have been reported to focus on armored reptiles and amphibians as prey," noted Emmons, pointing out that turtles, tortoises, caimans, crocodiles, and armadillos are common jaguar targets. This may help explain why jaguars are so often found along ocean beaches, deep within swamps, and upon the banks of waterways, where they have a long history of intercepting aquatic animals. Even horses, mules, and adult cattle are no match for *Panthera onca.* (The unusual muscle density of all felids allows them to kill and move animals bigger than themselves.)

Jaguars tend to develop specific hunting circuits, frequently following existing trails—or backcountry roads, where available—with all senses attuned acutely to the environment. (Young adult males—"dispersing transients" in the vernacular of biologists—may travel far to stake out new, unoccupied territories of their own.) While searching for a meal, a jaguar's eyes are wide open, hyperaware of any movement. Finely tuned hearing is attentive to the slightest noise. The cat's sense of smell monitors tiny nuances of odor detected in soil, foliage, and air.

Jaguars are opportunistic predators, stalking and ambushing healthy as well as old or disabled prey. Stealth, surprise, and lightning-quick reflexes are essential in order to achieve success. If it finds no feeding opportunities on its surveillance rounds, a jaguar may hide in thick brush and position itself to carry out an ambush. It could be on the lookout for an isolated individual, a juvenile, or—in the case of a herd animal such as a deer or peccary—a straggler left behind by disease, age, or injury. Reliance on this culling process, referred to by scientists as sanitation, varies among predatory species as well as among individual animals.

A jaguar preparing to attack is a study in extremely fine motor control, as it hunkers down with eyes wide and face forward in intense concentration. It edges ahead on cushioned footpads with robotic micromovements, getting as close as possible to its prey without being detected. To aid its final launch, the jaguar extends and bows its supple spine and firms its well-developed back legs and haunches. Finally, after the cat has judged the precise distance to its potential meal, it launches into a sprint, followed by a spring-loaded pounce before making a deadly bite. Sometimes a swift swat of the paw is employed to knock a target off balance.

"In sum," observed the *New York Times* science writer Natalie Angier in a 2003 report, "the jaguar has evolved a two-pronged approach to fetching dinner—stay virtually invisible to the last possible moment and then deliver an overwhelming blow."

Once its prey is captured and held fast by its sixteen-toe claws, a jaguar may use the dewclaws on the sides of its powerful forelegs to keep its victim enwrapped in a tight embrace. The sensitive nerve network around the cat's mouth relays precise information about what it is holding and biting, allowing the cat to make adjustments from first contact through coup de grâce.

Among the jaguar's keenest senses is its binocular eyesight, highly sensitive and with several times more acuity than humans' in dim light or darkness. The eyeball, pupil, and lens are proportionately larger in felids than in most other carnivores and are able to absorb much more light

Figure 3.1. The jaguar is distinguished from its close
relative the leopard by its stocky build, large head,
oversize paws, and unique fur patterns. Photo © Julie
Larsen Maher

than ours. Because the feline retina has more of the dim-light receptors
called rods than the cones used in bright light, it is largely color-blind. The
back of a cat's eye has a reflective layer called a *tapetum lucidum* ("bright
carpet"), which acts like a mirror in bouncing back light that already has
passed through the retina. In effect, this almost doubles the jaguar's visual acuity. It also is what gives such mammals the eyeshine reflected from
direct beams of light, including that thrown by flashbulbs. The jaguar's
exceptional nocturnal vision is particularly useful for hunting at dusk or

dawn, when the cat's dappled or dark fur serves as effective camouflage as it seeks the many animals active in crepuscular hours. Stripes and other black markings around a jaguar's eyes and mouth reduce distracting glare and may also promote nonverbal communication with other cats via specific facial expressions.

Jaguars, like dogs and many other animals, hear across a wide range of frequencies, including many above twenty kilohertz that humans cannot detect. The high-pitched rustle of leaves or of rodents squeaking, for example, may prompt a turning of the cat's *pinnae,* the rounded external ear flaps that home in like antennae on particular sounds. A complex set of muscles allows these and other felids to move each ear independently. (All cats also have the advantage of being able to turn their heads in almost a complete circle.) Using auditory nerves contained within two bulbous projections on the underside of its skull, the jaguar sifts through low-frequency noises for roars of other cats as well as felid growls, grunts, woofs, moans, pooks, chirps, and chuffs. The last of these is a blowing sound not unlike the release of pressurized air from a steam engine. All vocalizations made by jaguars mean something to other jaguars, though exactly what remains largely a mystery to us. This feline lingua franca is essential to survival, but we know only its basic vocabulary and grammar. Some observers believe that jaguars and other great cats may have individual voiceprints that allow others of their species to identify a sound as being from friend, foe, family, or potential mate.

All felid species share some essential behaviors and characteristics. A predilection for sunbathing is common among cats, for example. Another is the well-coordinated and agile use of unusually supple spines and flexible joints, along with voluntary control over hundreds of muscles. All felines seem able to spit, hiss, growl, snarl, and meow, but, depending on the bones and tissues of their voice boxes, only some can purr or roar. A mountain lion can purr but not roar, for instance, while the obverse is true for jaguars. When other demands don't intrude on a cat's schedule, simple relaxation often takes precedence. The naturalist Susan Morse suggested to Kevin Hansen in *Bobcat: Master of Survival* that a feline's five favorite activities often boil down to the following: "resting with its eyes open, resting with its eyes closed, lolling about, napping, and sleeping."

But eat alpha predators must, and these meat-eaters hunt very well. "The basic cat recipe is among the most adaptable and successful the animal world has ever produced," wrote Diane Landau in *Clan of the Wild Cats,* noting the remarkable bodies, refined skills, and effective behaviors of felids. *Panthera onca* is no exception.

Some researchers believe that the jaguar, relying on extraordinary spatial memory, notices even the most subtle or seasonal changes in its environment, including the habits of favored prey. It may remember, for example, at what time of year spider monkeys congregate around a particular fruit tree or when peccaries migrate through a specific savanna. Studies of these and other felids suggest that while a cat may not follow the same hunting circuit on a daily or weekly basis, it often revisits locations where it enjoyed past success. Based on camera-trap surveys, scientists know that jaguars can occupy territories of up to 525 square miles, depending on prey abundance and other factors, but may shift over the course of months or years from one part of their range to another. Drought, for instance, may cause such a predator to temporarily abandon areas where game is scarce, focusing instead on dependable water sources such as rivers and springs.

Unlike many cats—including other large felids of the Americas—jaguars are avid and accomplished swimmers who seem to genuinely enjoy being in the water. Streams and wetlands pose no obstacle. In coastal areas, jaguars even forage along ocean beaches for marine animals and waterfowl or paddle to islands if there is food to be had. But scientists are skeptical about long-circulating claims that jaguars dangle their tails in water to lure fish (rather than simply jumping in to grab them), or that this heavy animal leaps from high branches to ambush ground-dwellers. Jaguars nonetheless are very resourceful and climb trees easily. Once a kill is made, the carcass is often carried or pulled to a secluded spot for consumption. Jaguars do not routinely cover or bury carcasses for future meals, as do mountain lions and some other felines. The cats simply return at a later time or eat their fill in a single meal.

A jaguar seldom picks fights with other cats and is known to carefully mark its territory, in part to avoid troublesome encounters. The standoff of adult jaguars has been compared to the peaceful coexistence of nations that have atomic bombs but understand the cataclysmic consequences of actually using them. A jaguar is reluctant to risk being wounded in a warm, damp climate where deep cuts, damaged eyes, sliced tendons, and broken teeth can lead easily to impaired hunting, festering infection, accumulating parasites, or starvation. As a solitary animal—except when breeding or raising cubs—a wounded cat cannot easily care for itself. Too often, an injured jaguar becomes a dead jaguar in short order.

The jaguar's social-contact strategy, as with other cats, includes the leaving of tree scrapes, ground scratches, leaf mounds, scat (fecal matter), urine, anal-gland sprays, and oily body secretions in locations frequented by other felids. Odorific calling cards typically are left along trails as a way to maintain territories, reinforce status, indicate a female's sexual receptivity, establish travel corridors, and so on. Males shoot musky, chemical-rich liquids mixed with urine from their penises. These indicate approximate age, size, and other pertinent information. Females deposit their own set of information through secretions from their anuses and urethras. Glands in the cat's nose and mouth palate contain vomeronasal receptors that allow it to interpret such markings and deposits, which are as distinctive as a human face or voice. A highly specialized and sensitive nerve network, much more sophisticated than that of the human body, allows jaguars to parse and interpret the smorgasbord of smells left by others of its kind as well as different cat species.

Kevin Hansen, who has written books about the mountain lion as well as bobcat, told me a cat's aromatic deposits "are like bulletin boards." We discussed the phenomenon during a hike in southeast Arizona's Little Dragoon Mountains, an area that no doubt hosted jaguars in centuries past. "The olfactory sense of a large cat is so discerning that it can tell exactly who has left his or her mark at any given scrape or spray location," said Hansen. Pulling aside low-lying branches, he revealed deposits often made in the dry leaf litter below trees or shrubs, where liquids and smells can be absorbed and retained by organic material.

Figure 3.2. Scratch marks left by a jaguar on the trunk
of a forest tree. Such marks, along with scent sprays
and litter scrapings, are part of the sophisticated nonverbal network such cats use to communicate with one
another. Photo by Carolyn M. Miller

A jaguar has highly developed acumen when it comes to cheek oils,
footpad secretions, and other felid deposits. Urine and feces are the currency of common language in their world, giving otherwise solitary animals
a way to meet, breed, raise their young, and survive with minimal conflict.
On encountering a scent-marked site where feline fragrance is detected, cats
pull back their upper lips, bare their teeth, furrow their noses, and open
their mouths in a grimacelike spasm called flehmen. This allows airborne
molecules to come close enough to the palate and nose to sort out individual

smells. An olfactory gland called Jacobson's organ opens particular nerve pathways through sinus cavities as a way of "tasting" other cats. A male, for instance, may note the presence of sex hormones in a female's urine and oil secretions from her anus that suggest that she is available for sexual intercourse.

"The flehmen response is common to all cats," explained Hansen. "It is often interpreted (by humans) as a sign of displeasure, when just the opposite is often the case."

Like other felines, jaguars have glands between their toe pads that leave information-saturated oil residues behind, particularly when the cats broaden and stretch their toes in order to scratch. Other secretions produced by facial glands are rubbed against absorbent surfaces, often as a way of indicating territorial possession.

On the rare occasion when a pair of jaguars get together to mate, the entire process of courtship, foreplay, and intercourse may last as little as one hour. But the cats make the most of it. A receptive female may allow a chosen suitor to enter her as many as one hundred times in a frenzy of sexual bonding. Adult males have nothing more to do with a female beyond mating. A litter of cubs, clumsy and virtually helpless at birth, is born after a hundred-day gestation period. The mother must defend her offspring against potential predators, including adult male jaguars, disinclined as most are to share their domain with newcomers or to allow other males— even their own offspring—to disperse genes within their territories.

As elegant as a jaguar's physical construction may be, it is the animal's beauty and behavior that most of us find particularly arresting. In his photo-essay *Eyes of Fire,* Warner Glenn presents ten color plates from his 1996 Peloncillo jaguar encounter along the Arizona–New Mexico border. The rancher's images are startling and dramatic. The most compelling was made into a poster, a copy of which hangs in my dining room. It is a sidelong view of the mature male jaguar bayed by Glenn's dogs. The jaguar's chunky head is turned toward the camera, eyes glaring, its ears pulled back in an expression of defiance. The cat is crouched on a promontory beneath a milky sky, with a sweeping panorama of the parched San Bernardino Valley and green Chiricahua Mountains, a magnificent sky island that looms in the background to nearly ten thousand feet. A piñon pine obscures the tail and rear legs of the animal, whose muscular physique appears as tense as a coiled spring.

"The cat was not panting or out of breath," Glenn said, in describing the moment at which he took the poster photo. "He was quietly watching the dogs, then me, then the dogs again." The jaguar charged the picture-taker a few minutes later, but the attack was thwarted when a whirlwind of agitated tracking hounds jumped in between the two. Shortly thereafter, Glenn retrieved his dogs and the jaguar slunk away.

I have visited with Glenn and his wife, Wendy, several times over the years, and the couple's excitement about jaguars never seems to fade. At our first meeting I found Warner Glenn, as predicted, seemingly as tall as a saguaro cactus and as slender as a fence rail. His broad-brimmed hat, snap-button shirt, and rawhide complexion suggested a Madison Avenue version of a cowboy, but Glenn's appearance was as genuine as his firm handshake. "Lean and leathery as cowhide," the journalist Jeremy Kahn observed in a *Smithsonian* magazine article, "Glenn looks like he stepped out of a *Bonanza* episode." But behind the image "lurks a media-savvy and politically astute businessman." The fourth-generation cattle rancher is admired widely as a sharp-eyed hunter—and dedicated conservationist. He knew his jaguar pictures would stir debate.

In a four-page handwritten letter, Glenn told me he took his 1996 sighting "to mean that maybe we are doing something right" and that "just maybe jaguars can and are making a comeback." He speculated that the spotted cat felt at home in the Boot Heel and southeast Arizona because "he likes what he's found," a rugged place with plenty of game and few people.

Glenn told me that he and members of his family "continue to talk with other ranchers to convince them that we can live with the presence of jaguars and should be very proud to have them coming into the area again. It is a good sign of healthy ecosystem management on everyone's part." Maintaining a good prey base of deer, javelinas, and other wild mammals, Glenn wrote, "makes it easier for the rancher to accept the jaguars' presence, as he knows there will be something out there to eat besides the domestic livestock [a rancher] depends on for a living."

In 1997 the Glenns joined with dozens of others in the region in creating a quasi-governmental entity called the Jaguar Conservation Team as a way to study and protect the "northern jaguar," as biologists refer to the borderland version of *Panthera onca*. The "Jag Team" is a voluntary partnership led by the state wildlife agencies of Arizona and New Mexico. Participants include various government agencies, elected officials, nonprofit groups, scientists, professors, and ranchers. Its stated goal is to conduct research and coordinate conservation programs in the Southwest and adjacent states of Mexico.

    Critics dismiss the group—sometimes referred to as "the Jaguar Conversation Team"—as beholden to special interests, narrow-minded stakeholders, and political pressures. They note that it emerged during a highly polarized legal debate over whether the jaguar merited formal "endangered species" designation within the United States. (The Endangered Species Act, administered by the Fish and Wildlife Service, was passed in 1973 to protect creatures imperiled as a "consequence of economic growth and development untendered by adequate concern and conservation.") Once the cat was designated endangered, federal officials could conceivably have marked specific geographic areas as "critical habitat" necessary for the jaguar's recovery. They did not. Although *Panthera onca* won endangered status eventually, under pressure of legal action, the government has firmly resisted any suggestion that critical habitat be established or that jaguars be reintroduced to the United States in any overt way. At the core of their argument, federal authorities insist that such efforts are not necessary to ensure survival of the species because its strongholds have long been south of the border. This stance prompted some activists to give up on the Jag Team, suggesting aloud that if the United States is not willing to do more to protect its own jaguars, it has no business criticizing Latin American governments in similar positions.

"Obstacles to progress" on the north side of the border, the wildlife biologist Tony Povilitis told an interviewer in 2007, "include the reluctance by some folks to acknowledge the magnitude of the habitat loss problem, the myth that piecemeal or small-scale approaches alone can save the day for the jaguar, and the mistaken notion that real progress toward jaguar conservation can be made without the help of the Environmental Species Act."

But the Jag Team is viewed by other participants as an attempt to reach what one calls "the radical center." Rather than square off in divergent positions where consensus seems impossible, the group's leadership has voiced a desire to weave the art of compromise and skill of diplomacy to find common ground. To the extent it has succeeded, team efforts reflect the commitment of dozens of volunteer participants who meet several times each year near the border, usually in school auditoriums or city halls.

In late 1996 the newly formed Jag Team caught the attention of Alan Rabinowitz, the pioneering biologist of Cockscomb fame. Although Rabinowitz had spent much of the preceding decade studying the wild cats of East Asia, he remained an active jaguar researcher, making occasional forays into Belize, Panama, and other biodiversity hot spots. The sightings by Warner Glenn and Jack Childs, along with the dramatic photos taken by camera traps, provided evidence of activity among U.S. jaguars that Rabinowitz could not ignore. At the time he was a staff scientist in New York City with the Wildlife Conservation Society (WCS), which operates the Bronx Zoo and conducts wildlife research worldwide. Soon after the cats were seen in Arizona and New Mexico, Alan and his wife, Salisa, flew to Tucson for escorted field trips.

"Alan and Salisa got on mules and we took them up the canyon where we'd seen our jag," Jack Childs told me. "Alan was surprised that jaguars

Figure 3.3. The veteran tracker Jack Childs displays
2004 camera-trap photos taken by the biologist Emil
McCain of southern Arizona jaguars nicknamed Macho A and Macho B. Photo © Richard Mahler

could survive in such a rough environment, which he considered the extreme northern edge of the species' range."

Earlier in the week the couple had accompanied Warner Glenn into
the Peloncillos for a look at the ledge where the first cat had been cornered.

After their inspections, Alan and Salisa speculated that the Arizona jaguars were young males fanning out from larger breeding populations in Mexico. Such long-range dispersal is believed to be common when an area lacks sufficient prey to support a growing number of cats. This theory, widely held but still unproven, holds that older, dominant jaguars staunchly defend their established territories, while "teenagers" have little choice but to stake out new homelands.

Before returning to New York, Rabinowitz offered WCS support, including a grant to help pay for camera-trap surveys and a Southwest conservation plan for jaguars. As part of the latter initiative, Childs visited tropical Mexico and Brazil's Pantanal. His resulting guidebook, *Tracking the Felids of the Borderlands,* is now used by field researchers throughout Latin America. (After its initial support, WCS ended financial contributions to the U.S. jaguar conservation effort, apparently agreeing with the Fish and Wildlife Service that the Southwest jaguar had only a marginal bearing on *Panthera onca*'s long-term survival.)

Shortly before my initial conversations with Childs and Glenn, I sent an e-mail message to Rabinowitz seeking his cooperation and advice. We had both evolved professionally since our respective work in Belize, he as an action-oriented field researcher and me as a nature writer. More recently I had expanded my topic categories to include spirituality and art, while still reporting on the environment for National Public Radio and other outlets. Rabinowitz had spent considerable time in Thailand and Laos, where he trapped and studied endangered felids in national parks and sanctuaries. The scientist also traveled extensively in northern Myanmar (formerly Burma), where he helped to establish a large-scale reserve in the face of opposition by mine operators and big-game poachers.

"I am pleased to hear that your interest in jaguars continues and that you will be trying to help them in some positive manner," Rabinowitz replied, in a note sent from New York on his fiftieth birthday. "I would be happy to talk with you at further length." But that would have to wait, as the Myanmar project was taking Rabinowitz to Asia for many months. "Keep on trying to reach me," he urged, "and best of luck."

Notoriously forceful in his opinions and manner, the scientist had pushed hard for jaguar conservation during the eighteen years since he had left Cockscomb. He implied in interviews, including a national television appearance on NBC's *Today* show, that *Panthera onca* might be doomed to extinction without constant agitation and vigilance. "Even though jaguars were safe [in Belize back in 1986]," he told *Wildlife Conservation* magazine in 2001, "it doesn't take a brain surgeon to see the writing on the wall for the future. I think jaguars have a really good shot [at long-term survival, but] it's a race against time."

Rabinowitz repeated essentially the same message into 2008, when he left WCS to become president and chief executive officer of the Panthera Foundation, an international organization devoted to conservation of big cats that has subsequently taken over much of the jaguar research conducted previously by WCS.

CHAPTER FOUR

"We All Felt Really Blessed"

AN AMIABLE TONE PREVAILED at the first Jag Team meeting I attended. It convened in Douglas, Arizona, a former boomtown that faded with the closing of a copper smelter in the 1980s. By reliable accounts, the import of contraband and undocumented immigrants had replaced mining as the dominant industry. Immediately across the border was Agua Prieta, a much larger city considered a major drug-trafficking center. I stayed at Douglas's generic Motel 6 and walked to city hall, where forty-some in-terested parties were sprinkled among the soft chairs of a council meeting chamber. The proceedings were moderated by Bill Van Pelt, a shy, dark-haired biologist with Arizona's Department of Game and Fish. One of Van Pelt's duties was to report to the Jag Team about any purported Arizona jaguar sightings. He did so with aplomb.

"A resident on the outskirts of Tucson saw a dark shape moving through her backyard," Van Pelt began, reading from a list. "Investigator found no physical evidence and no photograph had been taken."

There was no shortage of sketchy reports. Several animals fleetingly seen were dogs or bobcats. Other phantoms could not be identified. When-ever there were fresh news stories about Arizona jaguars, Van Pelt chuckled, the number of jaguar "sightings" seem to skyrocket. However, at least one report was true. The usually formal Jack Childs, a sixty-two-year-old moun-tain lion tracker and land surveyor, could barely conceal his excitement in revealing that an unmanned border-area camera trap, triggered auto-matically by body heat in motion, had been tripped by a jaguar. The room echoed with *ooh*s and *ah*s as a color picture circulated. The nighttime flash exposure showed an adult jaguar striding purposefully across the desert.

The cat's markings distinguished it as Macho A, the nickname given to a jaguar photographed previously by Jack's remote-control equipment. His right-flank rosette pattern—resembling the face of the 1930s cartoon character Betty Boop—confirmed that he was not the cat seen by the Childs party nearly eight years earlier. That one, dubbed Macho B, carried rosettes on his right ribs that recalled the storybook hero Pinocchio. (*Macho* is the Spanish word for *male;* thus these jaguars were designated the first and second males photographed by Childs's project.)

Based on the unique pelage signatures, Jack Childs knew that Macho A was a cat his set-ups had photographed irregularly since 2001. But this would be his final appearance. Macho B had resurfaced after an eight-year absence. Pictures of the two males had been taken only four hours apart by the same camera. Jack speculated that Macho B may have been an archrival that had either driven away Macho A or killed him. Macho A was never photographed again.

The Jag Team meeting continued with questions and comments from various constituents. An environmentalist worried aloud that activities of the U.S. Border Patrol might disturb the ongoing transit of jaguars across the frontier. A sound engineer from Santa Fe proposed setting up loudspeakers in order to broadcast jaguar vocalizations—and presumably attract more cats. A third attendee inquired about research in Mexico, where scores of jaguars were believed to roam (and breed) in Sonora's backcountry. A fourth audience member, a red-faced woman whose family had raised cattle in New Mexico's Boot Heel for nearly a century, wondered what all the fuss was about: "We're nearly broke and barely hanging on; the last thing we need on our ranch is an endangered species."

After the assembly broke up, I approached Jack and Anna Mary Childs on the sidewalk, where a knot of jaguar enthusiasts lingered to pinch chewing tobacco and catch up on gossip. Decked out in cowboy hats, jeans, and boots, the couple had the friendly, down-to-earth manner of westerners who live close to the land. Jack's well-trimmed moustache, sky-blue

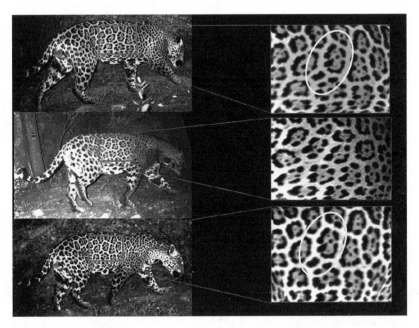

Figure 4.1. This camera-trap image composite shows how a distinctive rosette, resembling the cartoon character Pinocchio, confirmed that the much-photographed Arizona jaguar dubbed Macho B (top and bottom) is a different cat from Macho A (center). These male cats were photographed less than four hours apart by the same cameras, but their relationship is unknown. Photos and assemblage © Emil McCain

eyes, and gravelly voice recalled a middle-aged Paul Newman in the role of a straight-arrow sheriff. Anna Mary, a smiling strawberry blonde who let her wavy hair fall girlishly across her forehead, wore a colorful T-shirt emblazoned "Save the Tigers of Nepal."

The Childses agreed to join me for a late breakfast and to share first-hand details of their 1996 jaguar encounter. We strolled to a hotel named in honor of the Gadsden Purchase, whereby the land-hungry United States paid cash-strapped Mexico ten million dollars to acquire a railroad route through what would become southernmost Arizona and New Mexico.

❧❧

"My friend [and fellow mountain lion tracker] Matt Colvin and I were exercising our hounds," Jack Childs began in a firm baritone. "Anna Mary was with us, along with a taxidermist named Gavin Weller."

On the last day of August the group headed for an arid mountain range southwest of Tucson, not far from the border. Like the Peloncillos, about 150 miles east, the Baboquívaris are contiguous on a north-south axis with Mexico. Each range is an archipelago of flora and fauna untouched by recent ice ages. Unusual plants and animals, some found nowhere else, have retreated to the moist upper elevations of such ranges, providing a treasure trove of biodiversity.

With slight corrections and colorful nuances interjected by his wife, Childs told me that their trained dogs had followed a cat scent up a narrow canyon along an eastern slope. "Pretty soon the hounds got way up high on the side of a bluffy, steep mountain. We heard 'em jump the animal and bring it to bay."

Weller and Colvin were first to reach the treed quarry. Colvin was a fellow hunter, Weller an expert at mounting wildlife trophies. "My female dogs weren't acting as if they were on a lion scent," Childs recalled. He had noticed that the females, with more sensitive noses, were silent while the males barked furiously. "We could tell," he said, that the cat being followed "must have smelled different."

The Childses—who have trailed many mountain lions over the years —stayed on their mules and waited for Colvin and Weller to climb on foot through the dense brush and bring the dogs back. When the men did not return right away, Jack Childs set out to investigate.

"Halfway up the slope I met Gavin coming down. 'Matt sent me to get you,' he told me. 'We've got a jaguar.' I said, *'My goodness!'*"

Colvin began recording immediately with his video camera. The cat's sturdy legs and oversized paws dangled freely from a juniper branch barely strong enough to support its estimated 125 to 150 pounds. The jaguar peered down on the assemblage, probably the first humans the predator had ever seen at close range.

"We videotaped him for about half an hour," Jack recalled.

"The jag appeared to have just eaten a big meal and was pretty drowsy," Anna Mary elaborated: "If he didn't actually fall asleep on us, I think he surely was about to."

The time came when there was nothing left to do but leave, the lethargic feline still blithely perched in his aerie.

"We all felt really blessed," Jack concluded, with Anna Mary nodding at his elbow. "I never thought I'd see a jaguar. I figured it was just something you talked about around the campfire."

When asked, Jack agreed with others who had speculated that the jaguar might have come north from Mexico's Sonora under the protection of large borderland ranches, the nearby Buenos Aires National Wildlife Refuge, and the Tohono O'odham Nation. The reservation, adjacent to the west, covers an area as big as Connecticut but is thinly populated. A large cat could easily go unmolested there. In the weeks that followed Childs looked again for the animal, which he believed had headed into tribal lands. He found no sign of the jaguar.

By his own admission, the encounter changed this lifelong hunter's life. Soon after the sighting Childs quit his surveying job and dedicated himself nearly full-time to *Panthera onca* conservation. Over the next few years he befriended some of the world's foremost experts on felids. In 2001, donating their time and money, Jack and Anna Mary cofounded the Borderlands Jaguar Detection Project, devoted to the study of wild Arizona jaguars. In cooperation with ranchers, the U.S. Forest Service, and state game officials, the project set up scores of camera traps along the U.S.-Mexico border in what was deemed suitable jaguar habitat. Camouflaged cameras were attached to tree trunks along game trails that funneled into canyons or at pinch-points between boulders and other natural features. The units were checked regularly to determine what animals (including people) interrupted their infrared beams. Eventually, some digital cameras capable of recording live action were added, yielding territory marking and other distinctive behavior.

Besides relying on knowledge of the land garnered over more than thirty years of wildlife tracking, Jack used techniques calculated to snag a feline's curiosity and hold its attention. He sometimes rubbed a foul-smelling substance containing skunk urine about ten feet in front of the trap, hoping predators would investigate. (Calvin Klein's Obsession perfume, which contains a musklike secretion from the African civet cat, also has been effective in attracting jaguars.)

After my meal with Jack and Anna Mary, I decided to take a cursory look at the Arizona–New Mexico border region where jaguars were known to prowl. I had no illusions about my chances of actually finding a cat—and the Peloncillo jaguar photographed by Warner Glenn was long dead—but I wanted a firsthand glimpse of the Southwest's *Panthera onca* habitat. I had no intention of going to the exact spot where the 1996 jaguar was seen, only to the same general area.

Equipped with map, camera, sunscreen, water bottle, hat, and note-pad, I headed forty miles northeast of Douglas to Apache, a lonely inter-section marked by four houses, a school, and a graffiti-marred historical monument. I turned south for the fifteen-mile drive along a scruffy dirt road into Skeleton Canyon, an access point for a publicly owned portion of the Peloncillos. This dry wash was not far from where Warner had seen his jaguar eight years earlier.

At the canyon's mouth I was welcomed by a faded, bullet-shredded sign reminding me that the streambed was where, in 1886, U.S. Army General Nelson Miles accepted the surrender of the famous Apache war-rior Geronimo. Their meeting ended the last armed conflict with Geron-imo's band and virtually closed the chapter on three centuries of armed Native American resistance to European occupation of what is now U.S. territory.

For one who wishes to understand the elusiveness of the Apaches and the difficulty soldiers faced in capturing them, a day spent in Skeleton Canyon imprints a valuable lesson. This is challenging country, where run-ning water exists only during occasional summer cloudbursts and the sun dominates at least three hundred days of each year. The dizzying solar glare is relentless, and the nearly constant dry air sucks moisture from every inch of exposed skin. All surfaces seem to present razor-sharp edges and serrated, rotten rocks poked from every angle. The nettles, thorns, and needles of cacti, agave, sotol, and yucca made movement treacherous, even when traveling at the speed of a desert tortoise.

I was not prepared for a serious foray into the Peloncillo range, which loomed like a spike-pronged citadel, but I did spend several hours following the main Skeleton Canyon drainage and heading off-trail into side channels. I traversed the ominously named Devil's Kitchen, then up loose scree to the New Mexico state line, marked by a sagging barbed-wire fence. I topped a foothill ridge and took a few snapshots. Even on this late January afternoon, sweat dripped from my forehead. I was surprised that jaguars, which generally are observed close to perennial water sources, would frequent such a warm, dry environment.

En route I saw a couple of what I took to be bobcat tracks, along with those of rabbits and lizards. Mostly I saw telltale evidence of fellow humans—perhaps illegal immigrants or Mexican drug smugglers—mixed with signs of horses, coyotes, and cattle. I was surprised to find no hoofprints of javelinas (the regional name for the collared peccary) or deer, both preferred prey of the area's large cats. I climbed higher, anticipating that wetter, cooler elevations would reveal more wildlife. The mesquite thickets tore at my clothes, however, and the steep slopes became slippery in the absence of trails. I finally stopped, absorbed the palpable silence for a few delicious minutes, and headed back.

I threaded my rental car slowly along an unpaved track, stopping to open fence gates at regular intervals. (Private landowners later closed this road to the public.) It was hard to see how livestock could survive in such a sunburned place, yet I knew cattle—and cowboys—had roamed the Peloncillos for more than a century. Along the highway as I sped back to Douglas, I saw more evidence that *inmigrantes* had crossed surreptitiously from Mexico, probably using the same trails Geronimo had followed 120 years earlier. Human passage was marked by discarded water jugs, rusty tuna fish cans, spray-painted arrows, and makeshift campsites. In years to follow, aggressive security measures along the border would become a hot topic of concern among wildlife biologists and take legal challenges as far as the U.S. Supreme Court.

With peach light suffusing the pale western sky, I continued through venerable Bisbee, with its steep streets and open ore pits, then crossed the tree-lined San Pedro River before driving into the lavender shadows cast by high peaks of the Huachucas. A monstrous snowstorm was anchored above the sky island where I had met Mark Pretti three years earlier, and it taunted me with an unwelcome chill. I turned left near the army town of Sierra Vista before sliding through a blue-gray tunnel of darkness that concealed the artist-and-retiree communities of Sonoita and Patagonia. A drenching rain descended abruptly outside Nogales, a scrappy town straddling the border directly south of Tucson on Interstate 19.

I spent the night in a frigid motel room next to a truck stop where eighteen-wheelers idled impatiently, their drivers eager to reach the winter Mexican produce markets of Sinaloa. Ignoring the diesel rumble, I spread my paperwork and studied a map of the Baboquívaris, finally finding the narrow defile that had been the site of the Childs party's discovery. At midnight I switched off the desk lamp and lay down in darkness. My body was tired but my mind hyperactive. I wondered whether a hungry jaguar was somewhere out there, enduring the seasonal deluge that raged beyond thin sheetrock and aluminum siding. Perhaps the big spotted cat rested under a sandstone ledge or inside an abandoned mine shaft, waiting for an end to the tumult of thunder and lightning.

Instead of sleeping, I reviewed the day's events and mulled the invisible connection I felt with others who had searched for *Panthera onca*. Were we kindred spirits? Both Jack Childs and Warner Glenn hinted that they had always wanted to see a wild jaguar but had never dared believe it might actually happen. Their sightings seemed to have stoked passions now deeply rooted. How else could one interpret the extremes to which these men had gone in order to study and to preserve jaguars?

Warner had written a book that displayed handsomely the photographs he had taken during his 1996 Peloncillo encounter. The rancher had granted numerous interviews, taken an active role on the Jag Team, and helped convince skeptical neighbors that the jaguar was worth protecting. He spoke out against construction of restrictive fences along the international boundary that might block a jaguar's passage. He was a founding member of the Malpai Borderlands Group, a nonprofit venture that tried to close the gap between conservationists and ranchers, convinced that the two could not survive without mutual cooperation and respect. The collaboration had brought prescribed burns back to the region after eighty years of fire suppression and had restored habitats to protect other species, including the critically endangered Chiricahua leopard frog.

For his part, Jack had arranged for dozens of camera traps and other surveillance devices to be set up near the border, donating his time and skills while paying for related expenses out of pocket. He had studied big-cat tracking intensively, recruited scientists and volunteers to help scour Arizona in search of jaguars, and headed the Jag Team committee on depredation. Like Warner, Jack had given numerous talks, slide shows, and media interviews, patiently replying to the same questions again and again. Both were longtime mountain lion trackers whom urbanites might not expect to defend a predatory cat's right to exist. But the West was changing, and these outdoorsmen along with it.

The longer I reflected on the matter, the more appropriate it seemed that the lives of Warner Glenn and Jack Childs had intersected with that of the man most responsible for creating the world's first jaguar preserve, a scientist who remained a top authority on *Panthera onca*.

In 1983 Alan Rabinowitz was a bashful yet self-assured young man with a newly minted science doctorate and a deep love of animals. His particular fondness for four-legged creatures was spurred by an embarrassing affliction. A chronic stutterer through his teenage years, the twenty-nine-year-old Rabinowitz had discovered as a boy in Brooklyn that animals did not care how or even whether he talked. They did not laugh at him, hurl insults, or try to beat him up, as his schoolmates did. They did not assume that because of his body spasms and awkward speech patterns Rabinowitz was stupid, nor did they assign him to special classes for "retarded" children. The stutter was so severe that Rabinowitz's tense body would twist and twitch when he tried to speak—holding words within—and no amount of therapy seemed to alleviate the condition.

"It was easier to just not speak to people at all," he has said in describing his childhood. Instead, Rabinowitz would communicate with the turtles, rodents, and chameleons he kept in a bedroom closet. "[I'd] come home from school every day and yearn for the darkness and safety" of that refuge.

The strong-willed youth who adored critters managed to overcome much of his speaking difficulty by attending a special clinic during his first year of college. But Rabinowitz internalized life lessons that his nonhuman friends had taught him.

"Animals listened and let me pour my heart out to them," he explained in a keynote speech to the Stuttering Foundation of America. "At some point in my youth I realized that animals were like me, even the most powerful ones. They had no voice, they were often misunderstood, and they wanted nothing more than to live their lives as best they could apart from the world of people." The young man vowed to become a mouthpiece and advocate on behalf of his childhood companions, defending exotic and endangered animals that could not speak for themselves. The promise has served as a raison d'être throughout the researcher's career.

As a graduate student at the University of Tennessee, Rabinowitz turned his after-school hobby into a profession. By now he had added weightlifting and martial arts to his repertoire of interests, and his slight frame began bulging with muscles that carried him effortlessly through the Great Smoky Mountains. Dark-haired and broad-shouldered, Rabinowitz already had the square jaw and intense gaze of someone whose life's goal, in his words, had "shifted from wanting to be like other people to a determination that I would be better than anyone else."

Through happenstance, after doing radio-collaring and other fieldwork with black bears and smaller mammals, the doctoral candidate was invited by Wildlife Conservation Society cofounder George Schaller to implement a two-year field program on jaguars, then among the least studied of all large felids. Rabinowitz accepted the offer with enthusiasm.

"The first thing I had to do was find Belize on a map," he laughed. "Then I packed everything I owned onto my pickup truck and drove south."

Soon after arriving in the Maya Mountains, the newly minted scientist confronted a new set of communication problems. "Mr. Alan," as Belizeans called him, had difficulty convincing local residents of what he was doing. They were deeply suspicious of his explanation.

"Rumors soon started circulating as to why I was really in the Cockscomb," Rabinowitz recalled in his 1986 memoir, *Jaguar*. "Most [believed] my jaguar story was a cover for growing marijuana. . . . A few thought I was there to dig up Maya artifacts. When they realized that I was trying to actually catch jaguars—and what's more, not to sell them but to study them—they found [the notion] impossible to understand."

Rabinowitz selected a decrepit timber camp as his base, relying at first on past strategies used to study jaguars. He examined tracks and scrapes, hair and feces. While these methods provide valuable clues as to where jaguars travel, how they signal each other, and what they eat, they are no substitute for direct observation and monitoring via radiotelemetry. Once a cat is collared, radio signals allow researchers to learn specifics about its range and habits. Examination of a "live capture" jaguar also can yield blood samples, tooth condition, weight, approximate age, presence of wounds or parasites, and other data.

Rabinowitz was eager to follow as many of Cockscomb's jaguars as possible. He sedated and collared a few after placing chained live animals in cages and waiting for jaguars to entrap themselves. Hunting dogs and guides also helped with tracking and capture. Radio transmitters were attached to sedated jaguars, which were then followed electronically on foot or by airplane.

Some simple reasons help explain why scientific knowledge of jaguars was so limited in the 1980s, despite their being the object of curiosity for centuries. First, such cats are generally able to sense humans long before they are seen. Jaguars' senses are highly tuned, alert to the slightest unfamiliar noise, smell, or movement. Second, the animals are supremely well adapted and camouflaged for life in dense forests. A person looking for a jaguar will almost certainly be detected in time for the animal to take evasive action. Large cats, by their nature, strive to remain hidden and elusive. Over generations of natural selection, jaguars have refined concealment skills to a formidable degree. These factors, coupled with his troubled personal history, stirred mixed emotions in Rabinowitz.

"Here I was studying an animal because I loved it . . . and I wanted to protect it," the scientist recalled. "But in order to do that with animals like jaguars I had to drug them and put fairly sizable collars on them. Even if they were able to live out their lives, I had in some way handicapped them. I couldn't help but feel as if I was almost giving them a stutter." In the autobiography written at the behest of his friend the movie and stage actress Jane Alexander, Rabinowitz mourns the unintended loss of several study subjects, including the cat he called "my most special jaguar, [Ah Puch, which] died in my arms."

Though he remained passionate and unrelenting in his pursuit of his fieldwork, Rabinowitz needed help from the small number of Maya families still living in the Cockscomb. He knew that in order to conquer the tropical illnesses and parasites he endured—and to find jaguars efficiently—he had to adapt to the basin's human environment as well as to its indigenous creatures. The neophyte researcher resolved to explain his intentions more clearly. This meant broaching the controversial idea of creating a government-sanctioned reserve to protect jaguars. At first the resident Maya were dismayed at the prospect of giving up their comfortable homes, productive farms, and well-established hunting grounds. They did not want to leave the Cockscomb. When relocation was finally ordered, some Maya remained until forced by authorities to leave.

The residents were moved to the village of Maya Center, seven miles east of Quam Bank on the Southern Highway. Many of those relocated felt angry and bitter, but several obtained good jobs with the sanctuary while many others developed relatively lucrative forms of livelihood through eco-tourism. Besides restaurants, lodges, and tour services, their community became deeply involved in the production of traditional craft items, such as slate carvings, that sell at good prices to a steady stream of visitors who must pass through Maya Center en route to the Cockscomb.

Small vegetable gardens and fruit orchards now surround the homes of those relocated, who have found new places to hunt and fish. In addition, the village receives a portion of fees visitors pay to enter the park, which is overseen by the Belize Audubon Society. The long-term impact of such conservation strategies is unknown, and similar approaches in other locales have had mixed results. Managers of the Cockscomb, like other wildlife sanctuaries, are in an ongoing dance with surrounding communities about how best to steward resources they protect.

When Alan Rabinowitz began his fieldwork, only a handful of detailed scientific surveys of jaguars had been conducted in the wild. This made it almost impossible for governments and landowners to make informed decisions affecting *Panthera onca,* despite its mid-1970s listing internationally as an endangered species. Without useful data about the jaguar's territory, behavior, and needs, decision makers found it difficult to know how to deal effectively with such a large, wide-ranging animal.

Rabinowitz's mentor George Schaller, along with the Brazilian biologist Peter Crawshaw Jr., initiated some of the first comprehensive research on jaguars in 1977 and had limited success. They managed to radio-collar only two cats in a part of Brazil where hundreds were assumed to roam. At the dawn of the twenty-first century, however, the research climate had improved dramatically. In 2000 jaguar field studies were under way in central and southern Mexico; Caribbean Nicaragua; coastal Costa Rica; interior Panama; the llanos (grasslands) of Venezuela; the wetlands of Paraguay; and several ecosystems in Amazonian Brazil, as well as Arizona and its cross-border neighbor, Sonora. Dozens of scientists in ten countries were working to try to understand—and potentially save—the New World's largest felid.

An overriding concern among researchers was that jaguar populations were becoming dangerously fragmented. Unless prompt action was taken, scientists argued, groups of cats would become so isolated from one another that they might become inbred and die out. By the late 1990s WCS and one of its key financial backers, Jaguar Cars, had launched an ambitious five-year project with the goal of protecting contiguous habitat for jaguars from Mexico to Argentina. By establishing corridors to link existing jaguar strongholds, the theory was that the animal's genetic diversity would be preserved. As part of this initiative a major study of Cockscomb jaguars was conducted during 1999 by WCS, followed by large-scale surveys in 2002–2003 and again in 2005–2008.

In a testament to the ongoing challenges facing researchers, Rabinowitz declared in a 2003 interview, "Amongst all big cats, we [still] know the least about them." But some other felids, including the Asian cheetah and various leopards, may be gone before a real understanding of these jaguar cousins is achieved. "If we truly want to save [such] large predators," insisted Rabinowitz, "we're going to have to watch and protect them forever."

As I slid toward sleep in Nogales, my thoughts drifted back to an earlier trip to Belize, when I had hopped a Cessna for the short flight to Dangriga, landing on the dirt strip where I had seen the marijuana plane crash in 1987.

After unpacking and showering at the Pelican Beach Hotel, I headed for the same cocktail lounge where the innkeeper Therese Bowman had sketched on a soggy napkin a map of the then brand-new Cockscomb Basin Wildlife Sanctuary. Now married to the U.S. expatriate photographer and marine biologist Tony Rath, willowy Therese was as gracious as ever. She was eager to tell me about a scenic waterfall recently added to the local tourist itinerary, along with a newly excavated Maya ruin called Mayflower.

When I steered the conversation toward Cockscomb, which I planned to revisit the following morning, the innkeeper's face clouded.

"Have you heard the news about Alan?" she asked.

"Rabinowitz? No," I replied, "is something wrong?"

I knew that in the course of his full-bore research the zoologist's life had been threatened at various times by fevers, machete attacks, injuries, and accidents. I braced myself for the worst.

"Alan has been diagnosed with bone marrow cancer. It's a rare form of leukemia, slow-growing and in an early stage. The doctors say he could have another ten or twenty years to live, but they can't be sure."

I caught my breath and noticed a flutter in my stomach. Her anxious expression suggested that Therese felt something similar. The innkeeper and I were around Alan's age: me a little older, she a bit younger. In midlife it can be a shock to realize that one is vulnerable to fatal diseases. Youth's myth of immortality gets a rude debunking.

Sensing my discomfort, Therese reassured me that Rabinowitz still looked and acted like his vigorous, confident self. The scientist had kept up his weightlifting routine, for example, and remained an avid practitioner of aikido and tai chi. He was convinced he could defeat his cancer. "Alan's not a quitter," Therese declared. "He's still working as hard as ever, focusing as much as he can on getting well and tending to his family."

I knew nothing about Rabinowitz's personal life, so Therese quickly filled me in. He was married to a Thai woman, Salisa, whom he had met while doing fieldwork on the big cats of Asia. She was a medical geneticist, now with the American Museum of Natural History, who had been studying malaria. They were parents of two preschool-aged children, Alex and Alana. The family had a pleasant, two-story home in a suburb outside New York City. Physicians had said that the leukemia was in remission but might be reactivated by the stress of fieldwork in harsh tropical environments. For now, Rabinowitz was planning to scale back such activity. But as things turned out, not much—and not for long.

CHAPTER FIVE

"Well Drawn and Unmistakable"

BUENOS AIRES IS AN UNLIKELY name for the place. It translates as "pleasant breezes" and links in the public mind to the cosmopolitan capital of Argentina, famous for its tango and beefsteak. By coincidence, that waterfront city happens to be near the southern end of *Panthera onca*'s current habitat. Yet the Argentine capital looks nothing like the Buenos Aires National Wildlife Refuge, where seasonal winds often buffet dry desert air toasted to 110 degrees or more. The refuge, occupying much of Arizona's Altar Valley, is at the *northernmost* tip of the jaguar's range. Decidedly nonurban and noncoastal, it is a fragile, partially restored Sonoran Desert grassland, overgrazed in an earlier incarnation as a working cattle and sheep ranch. Surface water seldom flows here, but enough moisture and soil abide to sustain varied flora and fauna, including deer, javelinas, rabbits, and smaller mammals. As a result, the prey base for predators is solid.

I began this segment of my peripatetic jaguar search at the U.S.-Mexico border, where a narrow paved road passed by two inspection buildings and though a gated fence that marked the international boundary. The adjacent village of Sasabe, Arizona, was tiny. Fronting its main street were a few crestfallen houses, a modest post office, and a decrepit general store with the only gas pump for miles around. On the day of my visit a few people hung around with blank looks on their faces, as if their spirits had withered in the isolation and sun glare. Tall trees and a deep gully marked the center of Sasabe, betraying monsoon downpours of July and August.

Driving north, I passed numerous green-and-white U.S. Border Patrol vehicles and a mobile observation unit—resembling a prison sentry tower—perched atop a hydraulic scissors-lift. A uniformed sentry glassed the terrain for signs of suspicious activity. All territory immediately north of the international line was officially off-limits to unauthorized humans. Patrol officers here had their hands full. One of the nation's most isolated border crossings, Sasabe has long been a favored destination for smugglers of undocumented immigrants and illegal drugs. The location is considered ideal. Few people actually live in the Altar Valley, but the big cities of Tucson and Phoenix are only hours away by car.

At the time of my visit the 118,000-acre wildlife refuge, a short drive north of Sasabe, had been managed by federal authorities for nineteen years and seemed benign, almost friendly. Birds chirped, cacti bloomed, and smiling rangers struck up leisurely conversations. At the visitors' center I learned details from a video program about the big male jaguar treed and videotaped seven years earlier by the Childs party. Although the sighting location was pinpointed on a map, access to the remote mountainside was tightly restricted. Nevertheless, I wanted to at least get a firsthand look at the land a jaguar recently had called home.

I headed for the two primitive campgrounds closest to the site, several miles up a dead-end dirt road from the refuge's only paved highway. The area was unusually busy that winter weekend and all campsites had been taken. The occupants appeared to be sport hunters, a reminder that a key and sustaining purpose of U.S. wildlife refuges is to provide sufficient game for recreational harvesting and associated income to regulatory agencies. Armed Buenos Aires visitors sought deer, javelinas, and doves.

I drove farther north on a bumpy track and arrived shortly before nightfall at the mouth of a deep canyon. I parked and walked along a winding path into the foothills, only to confront a locked gate. A sign advised that the visiting public was allowed beyond the barrier only twice each month on guided tours. I cursed myself for missing one such tour earlier in the day. It would not to be repeated for two weeks. I then saw a secondary trail, abutting the fence line, winding down a steep slope. I followed it toward the streambed of the lower wash, which cut deeply into the shoulder of Baboquívari Peak, a massive wedge of volcanic rock that had been obscured by wispy clouds all day. Patches of snow clung to its near-vertical upper face and streaks of moisture-darkened cliffs just below the summit.

This sentinel is holy to members of the Tohono O'odham Nation, who believe Baboquívari Peak is home to their creator, I'itoi. A possibly apocryphal story tells of Tohono O'odham plunging from the peak during the colonial era rather than submit to Spanish rule. Today tribal members protest when climbers or Border Patrol agents get too close to the peak, as access is restricted to Tohono O'odham. One non-Native who reached the summit before this restriction was the then–chief justice of the U.S. Supreme Court, William O. Douglas, who ascended in March 1950.

As clouds cleared, clumps of evergreens could be seen crowding the mountaintop, with gravelly soil tumbling down a slope dotted by oak, piñon, mountain mahogany, and juniper. From a great distance came the squeaky-wheel bray of a feral burro.

Descending toward the siren call of running water, I was soon surrounded by lacy desert willows and white-skinned sycamores. The transition zone above the creek included thickets of mesquite mixed with barrel cacti, prickly pear, and a few saguaro. These last, majestic icons of the Sonoran Desert, stood up to thirty feet tall here and were easily two or three centuries old. There were other giants, too. One barrel cactus was level with my shoulder and covered with yellow fruit. Sprawling thickets of prickly pear were fortresses only a hungry javelina might penetrate. Owls appeared in the purple sky as other birds performed their twilight chorus.

I walked along the stream and checked for prints in every patch of moist sand. There was nothing extraordinary. The usual suspects—skunks, coati, and rodents—had passed by, but no cats. Such a waterway was a perfect corridor for a feline, and I was surprised none had left a calling card.

As I returned to my car in the gathering gloom, sheets of rain fell diagonally from a thundercloud moving swiftly westward across the Altar Valley, which had glowed bronze only a few minutes earlier. Fat raindrops splattered their roof-rattle on the drive back to the highway. A few creatures—rabbits and kangaroo rats, moths and bats—were illuminated by my headlamps.

I searched for a campground shown on my map, but became disoriented in the storm. Worried about getting a rental car stuck in wet sand, I settled for an unofficial site about twenty yards off the highway. Although it was early, I curled up on the back seat, clicked on the radio, and burrowed into my sleeping bag. A chilly wind picked up, and rain thrummed heavily before abruptly ceasing. I switched off the news from a Los Angeles AM station and lapsed into shallow sleep.

The pop of gunfire awakened me at midnight. I suspected that a few sportsmen had fanned out across the dark valley looking for game, despite sanctions against nighttime hunting. I wondered, "If someone spotted a jaguar, would it be shot?" There were severe punishments for this, including five-figure fines and jail time, but the temptation to bag such a trophy would be great. My thoughts drifted as I tucked deeper into my bedroll.

The car felt cramped and I slept badly. A baby-faced Border Patrol agent appeared just after four A.M., rapped on my window, and aimed his flashlight full in my face. The officer explained that a group of undocumented immigrants had been spotted moving north from the border. I was groggy and almost incoherent in my response. The man cocked his head and asked: "Are you all right, sir?" After I assured him that I was merely exhausted, he left me in peace. I hid my wallet and car keys, then slumbered a while longer.

Long before dawn I heard coyotes yipping far away. Otherwise the valley's solitude gave up only the stirring of dried mesquite pods. In the predawn chill I considered the ongoing human presence at Buenos Aires that might impinge on a Baboquívari jaguar. Here passed birders, hikers, botanists, ranchers, "anti-illegal" vigilantes, and game wardens, along with the aforementioned officers, Mexican immigrants, drug runners, and armed sportsmen. For the thousandth time, I marveled that even one jaguar had moved through this sparse landscape without being detected. Yet Macho B was verifiably real.

The snap of gunfire returned. This time it was close at hand, perhaps two hundred yards away. I considered my options. The only thing close to a weapon available was a two-foot tire iron in the trunk. I nixed the idea of retrieving it and relocated my camera beneath the rear seat. Seconds later, peering out the window with my head low, I watched a line of men in dark clothing running awkwardly through a stand of palo verde trees toward the highway. Each runner wore a bulky backpack and the kind of thick wool cap favored by skiers. I counted ten individuals, plus a man leading the group who carried no load. The leader gripped an automatic rifle and leaned forward as he ran stiffly toward pavement.

I was fully awake now, eyes wide and heart racing. My biggest fear was that the group would commandeer my car and use it for a getaway. It occurred to me that perhaps the shots had come from pursuing patrol agents and that I could be caught in a crossfire. But the running men, presumably Mexican drug smugglers, passed by without so much as a glance in my direction. I breathed a long sigh of relief.

Sleep was impossible now, but I felt it prudent to let time pass before making a move. At last a gauzy gray tinge grew in the eastern sky. An hour before daybreak, I turned the ignition key and headed back to the Baboquívaris, scattering jackrabbits and roadrunners along the rutted lane. A cold rain fell from an ominous sky sliced by a brutal wind. Within moments of my parking at the locked gate, icicle raindrops turned to slush. Black-bottomed clouds spilled over the ridgeline, and I cranked the heater to its highest setting. I had not foreseen such poor weather, and the wintry mix was getting worse. The upper elevation of Baboquívari Peak was now totally hidden. In an instant the wet snow turned powdery and began to accumulate. Even if I braved the elements on foot, the road might become an impassable mess. I could not risk such conditions in an Alamo-owned vehicle. It was time to abandon my brief quest for the elusive jaguar of the sacred mountains.

I did not resume my search for a full year. In the interim I compiled information about jaguars through interviews, videos, audio recordings, libraries, and every other resource imaginable: the Internet, newspapers, magazines, books, scientific journals, and official documents. Based on this sleuthing, I published several articles in print media and on the World Wide Web. My focus was primarily on the remarkable return of *Panthera onca* to the southwestern United States, although there were plenty of other news-worthy stories to tell. Continued poaching of jaguars south of the border, for example, was an important issue that received virtually no attention from journalists. Nor did the efforts of wildlife biologists to collaborate with ranchers, farmers, and government agencies in resolving human-jaguar conflicts. Also compelling was the setting up of sanctuaries and travel corridors between jaguar strongholds. Conservation of such large predators was a tricky task and some of the strategies being implemented were highly creative. But during this period polar bears and ivory-billed woodpeckers enjoyed their fifteen minutes of well-deserved fame.

As word of my project filtered into the close-knit community of *Panthera onca* aficionados, tidbits of information arrived from far-flung individuals: a former Game and Fish official in Montana, a whitewater river guide in the Grand Canyon, a biologist in Costa Rica, a researcher in California, a policymaker in Brazil, an undercover law officer in Arizona, and a journalist in Mexico. Others followed, filling in holes in my understanding of jaguars—and surprising me by revealing their own gaps. It seemed odd that several senior scientists in the field seemed unaware of what other experts, even those physically nearby, were doing. Before long, though, I realized that fresh information was sometimes slow to circulate simply because biologists were devoting as much time as possible to doing "the work"—in the field, lab, or office—rather than publicizing it. Or also because the time lag between field research and publication of results in a scientific journal was months, if not years.

❧❧

I repaired to the library and pored through old books, eager to learn what my great-grandfather's generation had known about jaguars. I read in a musty volume of Dodd Mead's *New International Encyclopedia,* circa 1906, that the jaguar range then was believed to extend "generally throughout South America" and north "possibly even into Louisiana and Arkansas." Only the highest mountain slopes and coldest tracts of Patagonia were said to be free of jaguars, which the encyclopedia labeled "the most interesting of all the wild-cats of the New World . . . [and] perhaps the most savage and intractable."

But things have changed in one hundred years. According to numerous sources, the modern history of exotic cats in Central and North America has been a litany of accumulating loss. By 2009 the jaguar has disappeared from an estimated 70 percent of the territory it inhabited north of Colombia at the time of European conquest. Much of this extirpation occurred after 1900 and accelerated with human population growth. South of the Panama Canal, jaguars are faring marginally better, but they are still losing ground overall. The greatest densities remain in the forested Amazon basin, Pantanal wetlands, Paraguay's Gran Chaco, and remote parts of Venezuela and the Guianas.

A jaguar homeland that extended in my great-grandparents' lifetime from Arizona's Grand Canyon deeply into Argentina is now fractured and poorly defined. One result of this muddled picture is that no one is entirely sure *where* remaining animals persist. "In truth," concluded Susan McGrath in a 2004 report for *Audubon* magazine, jaguars "are so widely dispersed and so secretive that nobody really has any idea how many exist throughout the Americas."

In 2009 educated guesses range between eight thousand and sixteen thousand cats: a variance of 100 percent. As recently as 2002 the World Conservation Union estimated that fifty thousand mature wild jaguars existed worldwide and gave the species a "near threatened" classification on the World Conservation Union's IUCN Red List. But by 2007 the decline of *Panthera onca* had accelerated, and the WCU concluded that "only the Amazon region provided a safe haven," noting that "the jaguar population was six to seven times less in Argentina, Paraguay, and Brazil compared with 1990." Many experts offer ten thousand as a realistic estimate of the 2009 jaguar population, while conceding that any number is impossible to verify. Fewer than three hundred jaguars are said to hang on in Argentina, South America's second-largest nation. Of Guatemala, one resident scientist told me, "There may be forty jaguars in the entire country, there may be four hundred; no one knows." The total for Mexico—once a bastion for *Panthera onca,* a place where hunting was sanctioned well into the 1970s—is probably far fewer than the two thousand cats projected in 2007 by one prominent Mexican ecologist.

Jaguars would be in worse trouble were it not for their highly successful predatory strategies, honed to a sheen over millennia. There are practical reasons for this. A meat-eating animal this big, with minimal fat reserves, requires many calories simply to survive. It must calculate carefully how much energy it will need to spend tracking and killing prey. Deer, peccaries, large reptiles, and monkeys are sensible choices inasmuch as the flesh of a single large animal can sustain a jaguar for days.

A jaguar has few natural enemies outside its own kind and will even sleep soundly in the open on occasion. A few animals—including the white-lipped peccary, crocodile, giant anteater, and anaconda—are known to be aggressive toward jaguars and occasionally kill them. But the cat enjoys considerable physical and sensory advantages over many other animals.

Citing one small example, jaguars depend on their vibrissae—the long, bristly whiskers that sprout around every felid's mouth—to stay well informed of their position in dark or tight locations. In a narrow space, these specialized hairs tell a cat how far it ought to proceed, since the longest whiskers are never wider than the skull, the widest point on a cat's body. (Their detached clavicle bones allow felids to squeeze through virtually any opening that is at least as wide as their heads.) The stiff, black-rooted shafts of cat whiskers are attached to sensitive nerves that, when stimulated, can provoke the animal instantly into full, watchful alert. These hairs are more effective dry than wet. Muscles at the base of the vibrissae can be used to bring the bristles forward or back in order to help cats "read" what they are up against. After an attack, the vibrissae also help such predators detect any lingering signs of life.

Another highly evolved survival strategy involves vocalization. Jaguars may announce their presence with a raspy roar. This series of low vibrations is more akin to an old man's phlegm-saturated cough or the bad muffler of a car than the proud proclamation of a master carnivore. The rumbling call originates when air passes from the lungs and vibrates tissue inside the jaguar's larynx, affixed loosely by cartilage to the hyoid bone behind and below the tongue. The roar often descends gradually in strength and loudness from start to finish, trailing away with a huffing or woofing sound.

A jaguar roar, which often accompanies hunting, can carry a considerable distance and draws the instant attention of any creature that hears it. Prey animals are extremely responsive to the sound, but other felids listen too. As a matter of survival, large cats have evolved to pay close attention to one another's physical spacing. Some researchers see this as a management strategy calculated to preserve dominance by specific cats in a particular geographical range.

It may be that vocalizations made by jaguars carry a sort of acoustic fingerprint unique to each individual, as has been theorized with other big cats. This is believed the case among tigers, for example, whose calls are known to have a low-frequency energy less distorted by vegetation and humidity than higher-pitched sounds. Indeed, there may be many specific vocalizations and sound signatures below the range of human hearing—known as infrasound—that large felids such as jaguars and tigers use to call their cubs, warn intruders, attract mates, and so on. This is another realm of zoological inquiry ripe for further exploration.

All cats belong to the order *Carnivora,* which evolved about sixty-five million years ago as dinosaurs disappeared and mammals proliferated. Cats are generally distinguished by short jawbones, forward-facing eyes, acute hearing, four stabbing canine fangs, meat-shearing carnassials (cheek teeth), and needle-sharp claws. They specialize in silently stalking or ambushing their prey, then dealing piercing death bites to head or neck during fierce, lightning-fast attacks.

Jaguars have had plenty of time to evolve their own particular specializations. Today's species is a holdover from a long-past era. According

Figure 5.1. Strong jaws, quick reflexes, and well-muscled bodies are among jaguars' assets in killing and eating well-armored prey, in this case an armadillo. Photo © Carol Farneti Foster

to professional interpreters of the fossil record, its direct ancestor, the long-extinct *Panthera onca augusta,* separated from others in its genus in North America between 600,000 and 1.2 million years ago. The modern jaguar's predecessor probably emerged first in Eurasia before crossing the Bering land bridge to present-day Alaska.

Panthera onca ranged throughout much of North America during the Pleistocene epoch, which ended with the close of the last major ice age about eleven thousand years ago. Fossils indicate that the cats once lived in present-day Washington, Idaho, Wyoming, Nebraska, Tennessee, Minnesota, and Pennsylvania. "Until the end of the Pleistocene," Ronald M. Nowak reported in *Walker's Wild Carnivores of the World,* the jaguar "occurred throughout the southern United States, and it seems to have been especially common in Florida." Indeed, some authorities believe that early jaguars thrived in each of the lower forty-eight states. Only the frozen north was beyond their ken.

During the Pleistocene the North American jaguar was more than twice as big as its modern descendant, or about the size of today's African lion. The 38,600-year-old bones of a specimen collected in an Oregon cave were those of a five hundred–pound jaguar boasting a fourteen-inch skull. Over time, as with bears, dogs, and camelids, jaguars made their way across the isthmus of Panama into South America. (Moving south to north, meanwhile, were armadillos, porcupines, sloths, and opossums.)

Jaguars once kept company with other large species that have since died out, including dire wolves, mastodons, mammoths, camelops, ground sloths, North American horses, North American lions, saber-tooth cats, North American zebras, ancient bison, and North American cheetahs. In fact, the jaguar is among few remaining *megafauna*—literally, "big animals" —that existed in great variety and number during the chilly, damp period that held sway before Earth's most recent set of ice ages. For reasons that remain unclear, most megafauna became extinct during the final ten thousand years of the Pleistocene epoch.

Mass disappearances before the Industrial Revolution are generally considered natural events, variously attributed to volcanic eruptions, epidemics, overpopulation, asteroid impacts, or climate change. But the extinction rate during the late Pleistocene was more than one hundred times what science has come to expect in similar situations. Because the disappearances coincided with the appearance of Paleo-Indians—meat-eating humans who competed with carnivores for wild game—archaeologists have joined the inquiry. They disagree on exactly when humans arrived in the Americas— though most believe it was less than twenty thousand years ago—but they advance the possibility that North America's early *Homo sapiens* contributed directly to the disappearance of at least some megafauna by killing and eating them. Like the virtually tame creatures of the Galápagos Islands, New World megafauna evolved without fear of being hunted by people. Climatic changes also may have contributed to the extinctions. The process may have taken generations, but eventually the more vulnerable large species died out.

The slow-moving ground sloth and the elephant-related mastodon and mammoth were among the first to go. The indigenous North American horse also disappeared, well before Europeans came to stay in 1492. The only camelid creatures that remained were the vicuña and guanaco of South America. (The llamas of today were bred originally from guanacos; alpacas are a cross of llamas and vicuñas.)

"At the time of these great extinctions, only between 10,000 and 50,000 human hunters were believed to be in the New World," write Joseph L. Chartkoff and Kerry Kona Chartkoff in *The Archaeology of California*. In at least two locales—prehistoric sites in Texas and Arizona—the skeletal remains of Paleo-Indians and jaguars were found commingled, confirming interaction between four thousand and eleven thousand years ago. Almost certainly there were more jaguars in those days than now, and far fewer people. It is conceivable that jaguars killed and fed upon human beings when opportunities arose. There is considerable evidence that the reverse also was true. It may be that then, as now, jaguars studiously avoided people. In the past few centuries, at least, they have been the least likely of all big cats to attack humans.

Reports of unprovoked attacks by jaguars are "extremely rare," according to the biologists David E. Brown and Carlos López González, and most are "nearly impossible to verify." Hunters, according to Alan Rabinowitz, state that a jaguar "often tries to escape rather than fight."

Fast-forward to the twenty-first century. In much of the continental United States, only the mountain lion remains a widely distributed top-of-the-food-chain predator. The grizzly bear and wolf were gone from many states by the early twentieth century and linger only in isolated pockets of the Lower Forty-eight. The range of mountain lions was reduced to remote parts of the West before a population rebound began during the 1980s. *Panthera onca* hung on only in the tiniest of numbers in extreme southern Arizona and New Mexico.

❧❧

I wanted to learn how jaguars might have been integrated into the lives of the indigenous people of the present-day United States. I was surprised, through interviews and library study, to find evidence of a rich, wide-

ranging relationship. Through parts of the Southwest where the cat was
well entrenched, for instance, evidence of habitation is found in ancient
murals and rock art left by the Pueblo and Mimbres cultures. These and
other Native groups made ceremonial items and trade goods with the
bones, teeth, talons, and pelts of the cats. Among present-day Pueblo
tribes in Arizona and New Mexico—as well as their widely presumed ancestors, the cultural group popularly called Anasazi—jaguar imagery appears in several *kiva* murals (wall paintings adorning sacred chambers).
Along the Río Puerco in central New Mexico, the Pottery Mound site (occupied from about A.D. 1300 to 1475) contains a spectacular painting of
a jaguar.

"This animal is well drawn and unmistakable, light brown in color
with dark spots," the ethnozoologist and rock art specialist Steve Pavlik
told me. "It is depicted reaching for a bird, possibly an eagle." According to Pavlik, the painting may relate to the Aztec eagle and jaguar warrior societies of central Mexico, a theory bolstered circumstantially by the
presence of parrot and macaw feathers at Pottery Mound. These and other
items of Mexican origin were popular trade items in the Southwest at the
time. (Looters destroyed a second jaguar painting at Pottery Mound before
archaeologists could examine it thoroughly.)

Another image of what may be a jaguar adorns a smoke-blackened cave
on the Pajarito Plateau of northern New Mexico, near the nuclear weapons
laboratories of Los Alamos. Around the time of the drawing's creation,
between A.D. 1325 and 1550, the area was occupied by a culture some social
scientists call Ancestral Puebloan that interacted with indigenous peoples
of Mexico. In the view of many anthropologists, Ancestral Puebloans relocated from drier uplands to the Rio Grande Valley during severe droughts.
Their descendants are said to occupy homelands (pueblos) in New Mexico
from Taos, near the Colorado border, southwest to Zuni, immediately east
of the Arizona state line.

Members of another long-enduring group, northeast Arizona's Hopi,
are also widely believed to be of Ancestral Puebloan heritage. They, too,
appear to have a *Panthera onca* connection. Several kiva murals found in

Hopi villages depict four-legged, catlike animals that some experts believe are jaguars. The anthropologist Leslie A. White theorizes that a supposedly mythical creature in Pueblo religion, the *rohana,* is actually a jaguar and may be the subject displayed in at least some of these murals.

Hopi hunters killed a jaguar near the south rim of the Grand Canyon during the winter of 1907–1908 and later displayed its pelt at their villages on Second Mesa, about one hundred miles northeast of Flagstaff. A photograph of what some believe is the same jaguar is reproduced in M. W. Billingsley's 1971 book *Behind the Scenes in Hopi Land,* in which the author claims that the cat's skin showed no arrow marks. "If true," reasons Pavlik, "this might reflect a ceremonial killing in which the animal was smothered with corn pollen" or dispatched in some other ritualistic fashion.

At Santa Ana Pueblo, twenty miles north of Albuquerque along the Rio Grande, White was informed by a tribal source that a rohana image unveiled there was a "big cat with spots." The anthropologist was further advised that it represented one of the "spirit hunters of the west" who, in turn, bestow their power on Santa Ana's adult males. A similar tradition exists at Zia Pueblo, a short distance west of Santa Ana.

Michael Flowers, a New Mexico State University anthropologist specializing in rock art, has identified a number of other centuries-old images made by indigenous people of the Southwest that he believes may be jaguars. These petroglyph cats are distinguished from mountain lions, according to Flowers, by spotting and banding drawn inside their outlines.

The northern New Mexico photographer Brad Draper has documented thousands of rock art panels and showed me an array of pictographs and petroglyphs in the western United States and northern Mexico that may be reasonably interpreted as jaguar imagery. A few are from south-central New Mexico's Three Rivers Petroglyph Site, where basalt rock faces showcase some twenty-one thousand glyphs and designs, including felids and other animals. Three Rivers is near the ruins of an ancient village of the Mogollon people, who flourished in the region from about A.D. 450 to 1400.

Less than a two-hour drive south, hidden in arid, desolate mountains above El Paso, Texas, Mogollon rock art depicting *Panthera onca* appears in a natural shelter known as Jaguar Cave and at a second location called Cave of the Masks. These intriguing artifacts suggest a Mesoamerican influence. The Cave of the Masks jaguar image, for example, wears a "shaman's cap" that some researchers believe reflects the cat's ritual function in an early civilization thousands of miles to the south.

The Hopi, Akimel O'odham, and Apache are among several tribes known to have prized jaguar skin for making arrow quivers, presumably to convey hunting prowess to those carrying weapons in such containers. For these and several other U.S. tribes within jaguar territory, it was believed that *any* object associated with this clever cat—from a likeness carved in jade to a tooth taken from a skull—accorded its owner special status and privilege.

Like the Apache, the seminomadic Diné (Navajo) were relative late-comers to the Four Corners, the area where Utah, Colorado, Arizona, and New Mexico meet. These Athabascans arrived from the north roughly between A.D. 1300 and 1500. Spanish colonial records from two and three centuries later report some Diné spoke to missionaries variously of a "meadow wildcat," "tiger," and "spotted lion." Each may have been a euphemism for jaguar. It also has been suggested that the same animal is among the "Cat People" referred to in Diné creation stories and may be the "Spotted Lion" that appears in some sand paintings, the symbolic designs used in tribal healing rituals. At least two ceremonies practiced by Diné medicine men and unrelated to sand paintings refer to spotted cats and their skins.

Like the Diné, the wide-ranging Apache came to the southwestern United States from the plains of present-day Canada shortly before European colonization. Throughout the Southwest, jaguars are said to have figured in hunting tales recounted around campfires by Apache and Diné warriors. Many members of both tribes were incarcerated during the 1860s by the U.S. military at central New Mexico's Bosque Redondo, where the cavalry officer John C. Cremony wrote that on his forays nearby "even jaguars were by no means uncommon." In a separate Bosque Redondo incident, a U.S. military scout filed a report describing an Apache renegade attacking an enemy with unusual ferocity. "I made jaguar medicine on him," the warrior declared, "and grabbed him like a jaguar and killed him. I was *like* a jaguar." This kind of close identification between jaguars and native people of the Southwest seems to cross strictures of tribe and time.

In southern Arizona, for example, a Tohono O'odham elder named Joe Joachim told the ethnozoologist Pavlik that his people "have always had deep respect for the jaguar and seldom hunted them. If a man killed a jaguar he risked contracting a sickness and would have to undergo an all-day ceremony. In this ceremony the medicine man or shaman used jaguar parts, including the skin and tail."

While researching a travel guide to the Palm Springs area, I found similar references to jaguars in a museum maintained by the Agua Caliente band of the Cahuilla tribe. For many centuries the Agua Caliente have inhabited the palm-shaded canyons of nearby foothills. A museum scholar assured me that the species played an important role in the spirituality of her forebears.

Jaguars, which the trapper John "Grizzly" Adams claimed to have seen in 1855 in southern California's nearby Tehachapi range, apparently were sacred to other bands of desert-dwelling Cahuilla, who reportedly did not allow the cats to be hunted or killed. But the jaguar's disappearance from the area around 1860 seemed to put an end to whatever related beliefs and rituals the tribe once actively maintained.

"In the 1920s," Richard Perry wrote in *The World of the Jaguar,* "W. S. Strong was told by the old chief of the Cahuilla Indians . . . that in the days of his youth he had been well acquainted with a large cat with a spotted yellow-brown skin and a long tail that lived in the mountains bordering the desert. His people were in the habit of following the tracks of this cat . . . in order to uncover the carcasses of deer buried by them; but the names jaguar and *el tigre* were unknown to him." (An inconsistent element in this story is carcass burial, far more common among mountain lions than among jaguars.)

The mammal authority C. Hart Merriam was told by an elderly chief of California's Kammei tribe that "a big spotted lion" was known to roam the Cayamacas Mountains near San Diego in the early nineteenth century.

Mountain lions continue to hunt deer in the Cayamacas range, while infrequent reports of jaguar sightings in the region, including adjacent Baja California, are invariably dismissed by scientific investigators.

Admiration of *Panthera onca* persists in Arizona, where a Yaqui dancer wore a jaguar mask during a 1997 ritual dance at the tribe's Tucson reservation. When asked by a TV news reporter why he chose to adorn himself this way, the dancer explained that he was honoring the return of a cat deeply revered by his tribe. The Yaqui man knew that the sacred animal had been glimpsed twice in southern Arizona—by Glenn and Childs—during the preceding year.

"The jaguar has a long tradition in Yaqui culture," according to Pavlik. "The word for jaguar in Yaqui is *topol*. There is a Yaqui village in Sonora [Mexico] named Topolovampo [place of the jaguar] in honor of this cat."

The jaguar's northernmost confirmed breeding population in northeastern Sonora happens to be near the confluence of the Bavispe, Aros, and Yaqui rivers, a source of precious freshwater for the desert-dwelling Yaqui. In several other northwest Mexico communities where indigenous cultures remain strong, jaguar dances are still held as spiritual rituals that pay tribute to the cat as a symbol of power and fertility.

It is likely that when English pilgrims landed at Plymouth Rock in Massachusetts and Spanish explorers founded Florida's St. Augustine, the "spotted cats" of Native American legend still prowled much of the Southwest. Sightings of jaguars—mislabeled as late as the 1920s in official documents as "American leopards" or "Mexican tigers"—were reported before 1850 by Spanish soldiers, missionaries, and explorers in California, Arizona, New Mexico, and Texas. Well into the nineteenth century, written accounts suggest that a few jaguars still roamed parts of western Louisiana and southern California.

Texas was also a nineteenth-century refuge for the cat. Among members of the Comanche tribe in Texas during the early 1800s the jaguar "was not infrequently met," wrote the naturalist Ernest Thompson Seton in 1920. The cat's skin, Seton noted, was favored by the tribe in making arrow quivers. While held captive by the Comanches during the 1830s, the white settler Dolly Webster of Texas related in her autobiography that she became fearful of a marauding animal she described as a "leopard cat." By 1910 jaguars had become scarce in Texas, where they virtually disappeared by the late 1940s. (One straggler was reported shot at Big Bend, near the U.S.-Mexico border, in 1962.)

In 1843 the explorer Rufus Sage described seeing "a strange looking animal . . . of the Leopard family" near Longs Peak in the Colorado Rockies. Jaguars in California, reported by Spanish and Mexican residents before 1826 in coastal mountains almost as far north as San Francisco, seem to have been scarce in the state decades before the last known specimen was killed near Palm Springs just before the Civil War. In New Mexico sightings were rare after 1910. Arizona was the last state known to harbor a breeding population of jaguars; females were hunted (and legally killed) as late as 1963. The last Arizona female was reported shot by a U.S. Fish and Wildlife predator control agent among the high-altitude conifers of the Apache National Forest. Six years later, jaguar hunting was banned altogether, although two cats were shot and killed after the law went into effect.

From the sixteenth through the nineteenth centuries, the jaguar appears in Southwest stories related by Basque shepherds, Mexican traders, French trappers, and U.S. mountain men. According to John Bakeless and other historians, the Spanish conquistador Francisco Vásquez de Coronado, while crossing New Mexico in 1540, heard about and possibly saw a large rosette-and-spot-dappled cat that at the time was said to prowl as far west as California, north into New Mexico's high plains, and east to Arkansas. While such accounts suggest a relative abundance of jaguars in the area during pre-European times, it is impossible to verify their accuracy. Anecdotes, most scientists agree, are the least reliable form of wildlife documentation.

Spanish colonials labeled the feline they encountered *el leopardo* or more often *el tigre,* mistaking it for the leopard and tiger they knew from Africa and Asia. Here, as elsewhere, colonists borrowed names based on what was familiar. We now know that the early Spaniards were partly correct. Although modern-era leopards and tigers are not native to the New World, molecular studies confirm that tigers are of the same genus, *Panthera,* and jaguars and leopards are cousins genetically.

Nevertheless, jaguars are a distinct species and differ somewhat from other great cats in appearance. Tigers are large—second in size among all cats to the dun-colored lion—and have pelage emblazoned with stripes rather than dots or rosettes. The leopard is built for speed as well as stalking, with a smaller frame, head, and paws than the stocky, low-slung jaguar.

"The jaguar can be further differentiated from the Old World leopard," according to the biologists David E. Brown and Carlos López González, "by the jaguar's shorter tail, barred chest, and the presence of solid dots *within* the broken rosettes."

Both species have short hair, a characteristic of cats that live in warm climates and therefore need to shed body heat quickly, but a leopard's rosettes are closed and do not contain spots. Leopards also have thinner necks and legs, perhaps because less brute force is needed to overcome their prey.

Intriguing but dubious claims of jaguar, leopard, and tiger sightings surfaced in British colonies before the American Revolution. Some accounts placed big spotted cats in Appalachian forests during the eighteenth century. A 1711 summation from the southeastern mountains, shared by John Lawson in his *History of North Carolina,* notes that "Tygers . . . are more to the Westward. . . . I once saw one that was larger than a Panther, and seemed to be a very bold Creature. . . . It seems to differ from the Tyger of Asia and Africa." Another writer of the same period insisted that jaguars were encountered in the mountains of North Carolina as late as 1737. Richard Harlan, a contemporary of the famed nature illustrator James Audubon, wrote that jaguars were seen east of the Mississippi into the early 1800s, while the French naturalist Constantine Rafinesque made similar claims during the same era about "the large wandering Tygers or Jaguars of the United States." According to records in the Smithsonian Institution Archives, Rafinesque claimed that he saw jaguar skins nailed to the walls of barns along the U.S. frontier and cited instances in which hunters said they had killed jaguars in Kentucky, Ohio, and the Lake Erie region of New York. The Frenchman theorized that the cats migrated north and east in summer months, retreating south during winter.

Jaguars aroused plenty of interest among the first European arrivals in Mexico's arid northwest. The German-born Jesuit priest Ignaz Pfefferkorn, who served rural missions in the region from 1756 to 1767, made detailed observations in his 1795 book *Sonora: A Description of the Province.* "Tiger skins are prized above all others by the Indians," he reported,

> who use them for fashioning handsome quivers. The quest for such ornaments inspires the Indians with great daring. They search indefatigably for the track of the tiger, pursue the animal until they come upon it, and slay it with their arrows. At times the tiger gives them much trouble, for he may put up a desperate resistance or he may hide himself in a cave to escape death. The Indians are not stopped by this. They light a fire at the cave's entrance and thus force the animal to emerge. Upon its appearance, it is met with arrows, and rarely does the first shot fail to slay it.

In other writings, Father Pfefferkorn elaborated on jaguar hunting by indigenous people in South America, apparently summarizing reports from other colonials he met. The missionary paints a dramatic picture:

> [In the tropics] a sharp knife, three spans long, and a wooden spear, the point of which has been fire-hardened, are the weapons with which they attack the tiger. The spear is carried in the

left hand and the knife in the right. When the animal has been cornered and brought to bay, the hunter offers it his left arm, which has been heavily wound round with rags so that the claws of the animal cannot injure it. The tiger instantly seizes the arm with his paw and in the same instant the [hunter] strikes off its paw with the knife. The raging animal then tries to take revenge with the other paw, but the [hunter] just as dexterously renders that one useless as well. Thus the tiger is disarmed and killed without further danger.

Pfefferkorn's colorful accounts exemplify various early writings by Europeans that were at best misleading and at worst inaccurate in describing jaguar temperament. Though a savage fighter when threatened or trapped, *Panthera onca* is not normally aggressive toward people. The actual risk of death by jaguar is minuscule. In the course of my research, I found only three verified cases of jaguars killing humans during modern times. A fourth zoo attack, in 2009, was nonfatal.

In 1985 a two hundred–pound, six-year-old jaguar attacked and killed a pregnant zookeeper, Gayle Booth, at a Michigan facility. According to a wire service report, the neutered male "acted instinctively" when it managed to open an automatic door in a holding area at the John Ball Park Zoo in Grand Rapids. Following an investigation, officials decided that the killing was an accident and did not destroy the cat. In another fatal incident, this one in Austria, a jaguar attacked a zookeeper who had not followed proper safety protocols. This cat, too, was not found at fault for its behavior.

On February 24, 2007, a 140-pound male jaguar named Jorge mauled a zookeeper at the Denver Zoo. The employee was attacked in a service hallway adjacent to the felid's enclosure, to which a door was found open. Ashlee Pfaff died soon after the attack, and Jorge was killed when he approached workers trying to save their twenty-eight-year-old colleague. A necropsy found no incriminating evidence, and the animal's history suggested nothing abnormal. Pfaff's neck had been broken, but, because there were no witnesses, what prompted the killing remains unknown. The Colorado facility, like all accredited zoos, requires training of its workers in strict and precise safety protocols. When such procedures are not followed with captive jaguars, perhaps through overfamiliarity on the part of humans, the risk increases that the animals will turn on their caretakers. Such is the case with all large cats, as illustrated tragically when a white tiger owned by the magician and animal trainer Roy Horn attacked and critically injured him during a Las Vegas performance in fall 2003, and four years later on Christmas Day when a Siberian tiger at the San Francisco Zoo mauled three young male visitors, one fatally. Press reports quoted eyewitnesses as saying the men, said to be either drunk or high on drugs, taunted the cat before it leaped across a below-recommended-height barrier and attacked them.

In 1972 the naturalist A. Starker Leopold wrote in *The Wildlife of Mexico* that he was unable to authenticate any report of a Mexican jaguar ever becoming a man-eater. "Men undoubtedly have been killed by cornered or wounded jaguars," wrote Leopold, "but unprovoked attacks are rare. In this respect the jaguar differs from its relatives in the Old World. Lions, tigers, and leopards may all become confirmed man-eaters under certain conditions; the American cats fortunately do not."

With the passage of time, Leopold's assertion about at least one "American cat"—the mountain lion—must be challenged, though fatal attacks even by these felids remain rare. About twenty-five kills of humans by mountain lions were confirmed in the United States and Canada between 1900 and 2009. This compares with the twenty or so people killed in the United States *each year* by domestic dogs. Similarly, gray wolves are widely feared in North America even though the animals made no documented fatal attacks on humans during the twentieth century.

Peter G. Crawshaw Jr., a Brazilian wildlife biologist and carnivore expert who did groundbreaking research with the naturalist George Schaller, told me in an e-mail interview that he knew of only a handful of situations in which jaguars are said to have made unprovoked attacks on humans. "One was in Misiones, [northeastern] Argentina," wrote Crawshaw, who emphasized that these accounts could not be verified. "This guy survived. Another was near Cáceres, in Mato Grosso [southern Brazil], which was fatal."

The scientist went on to relate a story told to him by an Italian physician who had worked in the remote village of Catrimani, in northern Brazil near the Venezuela border. The doctor claimed that jaguars consistently preyed upon residents, most of whom were indigenous Brazilians. The man showed Crawshaw a photo of a victim who reportedly had been snatched from a hammock by a jaguar as he slept. "The man had been severely bitten through the head and never quite recovered his mental faculties," according to the informant, who was not a witness. "This was not an isolated incident," the Italian continued. "When this [kind of attack] happened, the men in the tribe used to follow the jaguar until it was killed, to avoid further threats from the animal."

During early days of settlement, colonials decimated jaguar populations in much of South America. This was accomplished largely through the use of firearms, horses, and hunting dogs. Spanish and Portuguese authorities from New Mexico to Argentina put bounties on jaguar skins, and hunters were eager to collect. "The real conflicts between humans and jaguars" started "in the mid-sixteenth century with the introduction of domestic livestock," Alan Rabinowitz has noted. "The reaction to such conflict has always been to kill jaguars."

My forays into the borderlands of Arizona and New Mexico left me with mixed feelings. I was disappointed not to find the slightest trace of a wild jaguar, yet I was pleased to meet a wide spectrum of fascinating people, from biologists and environmentalists to ranchers and hunters. They appeared strongly committed to helping this endangered cat prosper in the American Southwest and the adjacent Mexican states of Sonora and Chihuahua. I was eager to spend more time with these folks, but first I wanted to head south into tropical Mexico and Central America, where jaguars and their legacies were more abundant. In order to better appreciate these elusive creatures, I wanted to share their world—and to meet the field researchers, rural residents, government officials, zookeepers, and conservationists who keep them company.

CHAPTER SIX

"The Model for How to Live"

THE WAY GUILLERMO MORALES tells his story, a little dog deserves full credit for the discovery of a lifetime. Without Perrito's pursuit of a nondescript Central American jungle rodent, the priceless treasures of Che Chem Ha might have remained sealed for another thousand years.

"Perrito was running fast after the *tepezcuintle* and then—poof—he was gone," recalled Morales, slashing his hand through the air like a machete. "I had no idea what happened to him."

We were standing on a well-worn, muddy trail near the slate-gray Macal River, a few miles from the Guatemala border. This remote corner of Belize was opened to settlement during recent years as two controversial dams were built on the Macal's headwaters. These hydroelectric projects, upstream from a roaring cataract called Vaca Falls, created new roads seized upon by land-poor Belizeans as well as eager homesteaders encroaching illegally from Guatemala's adjacent state of Petén.

Morales, a native Belizean with a wispy moustache, soulful eyes, and solemn demeanor, walked to a car-sized shrub at the base of a near-vertical incline. "This bush was here—and when I looked behind its branches, I saw Perrito go into a small hole between two big rocks." The frightened tepezcuintle managed to avoid the clutches of the determined dog, which emerged dirty but unscathed a few minutes later. Morales noticed a slight draft of cool air blowing from the ragged opening. He knew that this was a cave, common in the limestone hills.

"It was late afternoon and the hole was too small for me to go inside," explained the farmer, only seventeen years old at the time of the incident. "So I came back the next day with a pick and shovel." The enlarged entrance led to a passageway that widened into dry chamber. From here, a second tunnel snaked deep inside the mountain. "I started finding pieces of broken pottery on ledges high above the floor," Morales said, in a whisper of astonishment. "Then I found big clay jugs—and lots of other things."

At first Che Chem Ha Cave was a well-kept secret, known only to the Morales family, whose homestead is a short walk downhill. The opening was expanded by Guillermo and later secured by a locked grate that has dissuaded most (but not all) would-be looters. Eventually, the family decided to go public. Using flashlights and fluorescent tape, Morales started guiding select visitors through the maze. He instructs them to move with extreme care through grottos damp with humidity and slippery with mud, pierced by stalagmites and stalactites. In some places the cave pinches and staggers in a way that demands use of ropes and handholds. The payoff, as I discovered during a ninety-minute tour led by Morales, was beyond imagining.

On shelves along the walls of its larger rooms, Che Chem Ha holds more than 110 earthenware pots, the majority larger than punch bowls. While most are plain, an exquisite few are painted with now-faded pigments. One image is clearly that of a bird, another is a figure that could be a small man or a monkey. Many vessels either are cracked or have holes in them. Deliberate damage is expected in Xibalba, the sacred Maya underworld where priests and rulers were consigned after their deaths. The oversized polychrome containers were meant to carry human spirits into Xibalba, and intentional breakage at burial ensured that the phantoms would not be carried back to an uncertain fate in the land of the living. Some ceramics are placed upside down, others stand upright with fitted lids. A few, deposited for ritual use by departed royalty, contain offerings of corncobs and annatto seeds. (Annatto is a spice and reddish coloring agent from achiote trees, native to the region.)

During my circuit through this labyrinth, slick and clammy with stale humidity, I learned of other artifacts discovered by the government archaeologists who studied Che Chem Ha in 1989. The inventory included nine human graves as well as wall paintings, incense burners, obsidian blades, stone tools, and—at the heart of the cave—a circle of rocks around a tall limestone stela, where animals were ritually sacrificed and prayers chanted during gatherings of Maya elite.

"I would like us to turn our flashlights off," Morales said, when our group reached the eighty-foot-high chamber dominated by the roughly carved stela, its base blackened by primeval fires. "Let's sit in silence at this altar and pay our respects to the spirits that remain. This was a church, you might say, for my ancestors." Morales extinguished his light.

The darkness was thick and velvety, as palpable as wool. When I passed my hand in front of my face, I was blind to shape and motion. A frisson of acute awareness prickled my brain. Was it my imagination, or did I sense the psychic energy of departed souls? Suddenly, I was alert to the slightest vibration, including the beating of my heart and the rhythm of my breath. The fine hairs on my arms became erect, sensitive to a dozen exhalations around me. A quiet calmness moved over us gradually like a warm, soft blanket.

⋙⋘

"How did you feel?" Morales asked, switching his flashlight back on.

"Completely safe," I murmured, "and fully alive."

Morales smiled his approval. Every member of our group seemed moved, as if we had experienced a small miracle.

Within Che Chem Ha, I found out later, those buried were assured of eternal protection by jaguars, the godlike guardians of portals to Xibalba. This seems to have been the belief in all sacred Maya caves dating from a particular era. The Morales site is one of many in the lowland Maya homeland where discovery of artifacts has contributed to a new understanding of the jaguar's status in the imagination of New World peoples. Shortly after my visit, a Belizean archaeologist informed me that previously unknown caves—many with artifacts—are still being found throughout his country.

"In some of these chambers," the archaeologist said, "we have to work our schedules around those of jaguars. They continue to use the caves as sleeping quarters or as shelters in which to rear their young." My eyes widened at the prospect of stepping into a cave and confronting a big cat.

Figure 6.1. Jaguars use caves as shelters and den sites.
The ancient Maya of Mesoamerica deified the cats as
protectors of Xibalba, the sacred underworld. Photo
© Carol Farneti Foster

"So far," the archaeologist assured me, "we've accommodated one another."
(This type of behavior is confirmed in *Walker's Wild Carnivores of the
World* by Ronald M. Nowak, who notes that a jaguar "may den in a cave,
canyon, or ruin of a human building.")

My interest in jaguar veneration parallels an enduring fascination with
indigenous cultures of southern Mexico and northern Central America,

the geography referred to as Mesoamerica by social scientists. I first traveled south of the border as a small child. My mother's eldest sister—Aunt Lucie—had a Spanish colonial–style home perched on the Texas bank of the Rio Grande with a clear view across the water into Nuevo Laredo. Her late husband, Jack, had been a journalist and public relations man who did considerable business south of the border.

At age eight I was thrilled to walk midway across the International Bridge and stand above the swirling chocolate-brown water: one foot in the United States, the other in Mexico. It was exciting to be in such dissimilar worlds simultaneously. On the U.S. side, English was widely spoken and street life was mundane. In Nuevo Laredo, any kid my age could stroll into a bar, buy a firecracker, or openly smoke a cigarette. Painted prostitutes trolled for clients on the sidewalks, and men with bootblack pompadours hawked "miracle cure" goat gland extract atop soapboxes in the plaza. Nothing seemed forbidden. Back at Aunt Lucie's spacious hacienda, situated mere yards from a border that in those days was as easy to cross as a city street, she showed me artifacts picked up (no doubt illegally) by her husband at Maya sites in northern Guatemala and southern Mexico. I still keep a corn-grinding stone on my nightstand that Jack collected at Yaxchilán.

An affinity for Latin America has never left me. During the 1970s I learned basic Spanish and worked extensively with Mexicans and Mexican-Americans at a public radio station. For three years I produced bilingual programs for *campesinos* (farmworkers) and trained teenagers to become *locutores* (announcers). Beginning in 1986 I spent weeks on end traipsing around Central and South America. During these visits I was drawn to museums, galleries, and ruins that displayed artifacts of early civilizations. In Ecuador I saw vestiges of the ancient Incan culture. Farther north, I wandered through the monumental Maya ruins of Palenque, Caracol, Chichén Itzá, and Tikal, as well as the simple farming villages of Guatemala's highlands and fishing camps of Belize's Turneffe Islands.

I noticed early on that the jaguar played an important role in virtually all Mesoamerican cosmologies and cultures. While the cats were rarely encountered in the flesh, their images were—and are—seemingly universal.

Throughout the region, jaguar images and references occur in everything from business names to song titles, flashy advertising to fine-art paintings, ancient ornaments to modern toys. As interpreted by humans, a flesh-and-blood animal has mutated into thousands of inanimate forms. I wanted to find out how long this had been going on—and why.

❧❧

Few wild animals have engaged the human heart, psyche, and soul as deeply and as consistently as have jaguars. For at least fifteen thousand years—and possibly much longer—this charismatic cat has wended its way into religions, cultures, and art. A creature of great imaginative allure, it has symbolized deities and natural phenomena, virility and power, royalty and magic. The jaguar has been associated alternately with healing and destruction, darkness and light, versatility and single-minded focus.

While all of the world's great cats—particularly Africa's lions and Asia's tigers—have inspired legends and cosmologies, only the jaguar has infused such a wide and diverse spectrum of beliefs and practices in the Western Hemisphere. From California to Patagonia, the species long has been assigned a high rank in the breadth of societies large and small.

But why? Social scientists offer various theories. Among many indigenous peoples of the Americas, the jaguar, according to the archaeologist Michael D. Coe in *The Cult of the Feline,* "represents an energetic principle, the natural life force that, on a social level, has to be controlled if moral order is to be preserved." Given its status as the top predator in virtually all of its range, the cat enjoys a special place in the minds of Latin Americans even today. Observes Mark Miller Graham in *Reinterpreting Prehistory of Central America:* "In many contemporary New World tropical societies, the jaguar is a male symbol imbued with connotations of leadership, sexual prowess, and dominance over females."

In the larger scheme of things, to cite a parallel example, it may be that the jaguar plays a role very similar to that of the world's large bears. According to the social scientist Joseph Campbell, the biggest bears of ancient Europe, Asia, and North America may have been the first wild animals to be widely venerated and mythologized by humans. "The [most powerful regional] animal becomes the model for how to live," Campbell told the journalist Bill Moyers in the 1986 PBS-TV series *The Power of Myth*. "It is superior, and sometimes [such an] animal becomes the giver of a ritual."

So it seems with jaguars, which historically have overlapped with large bears (the grizzly) in only a small part of the cat's habitat (Mexico's northern Sierra Madre and a few mountain ranges in the American Southwest). With no real terrestrial challenger, it is not surprising that admiration, respect, and religion are entwined throughout the complex and enduring human relationship with *Panthera onca*. Few animals of such strength have coexisted with us for so long, which in itself is a remarkable feat.

"Hostility between people and carnivores may be deep rooted," speculates biologist Nowak. "The diets of jaguar, puma, and tribal people in the neotropics overlap widely and . . . therefore we humans have always been in competition with [these] big cats."

Because of its obvious hunting prowess and superior sensory abilities, the jaguar also has been a "totem" or "power" animal for humans. In simplest terms, this means that characteristics of the creature are believed to be transferred to those closely associated with it. This can happen by means of rite, deification, or the wearing or display of parts of an animal's body, such as fur, teeth, claws, or bones. When the real thing is unavailable, a mask or carved likeness will do.

"Around the campfires of Mexico," A. Starker Leopold writes in *Wildlife of Mexico*, "there is no animal more talked about, more romanticized and glamorized, than *el tigre* [the jaguar]. The chesty roar of a jaguar in the night causes men to edge toward the blaze and draw *serapes* tighter. It silences the yapping dogs and starts the tethered horses milling. In announcing its mere presence in the blackness of night, the jaguar puts the animate world on edge."

In his book *Monster of God*, David Quammen speculates that European colonials launched deliberate campaigns to exterminate "big, flesh-eating beasts" that not only offer mortal peril to settlers but "hold pivotal significance within the belief systems of the natives." The underlying

Figure 6.2. This Mexican *vaquero* near San Bernardo, Sonora, wears jaguar-hide chaps, circa 1908. Some contemporary cowboys in Mexico shoot jaguars out of concern that the cats will prey upon their cattle. Courtesy Arizona Historical Society/Tucson; image #62002

philosophy, he wrote, boiled down to this bloodthirsty strategy: "Kill off the myth-wrapped tiger. Kill off the lion. You haven't conquered a people, and their place, until you've exterminated their resident monsters." Like a good shepherd guarding a flock of sheep, the benevolent ruler was presumed to eliminate alpha predators not only as a means to protect livestock but to control the people he ruled.

"The jaguar's hold over men's minds and belief far surpasses in its universality that of the tiger and the inhabitants of India," writes Richard Perry in *The World of the Jaguar*. "The universal dread of the jaguar is, and always has been, predominantly of a mystical nature . . . of an animal possessed as of supernatural powers."

Like the bear, eagle, wolf, lion, crocodile, tiger, and other dominant carnivores, the jaguar moves easily and without fear, sufficient unto itself. A jaguar not only has estimable physical strength, it exercises its authority to full advantage in any situation. Partly because of this, jaguars are the envy of those humans who seek to manipulate other animals, including their own kind. Through no ambition of their own, these felids have been associated with certain cultural values judged desirable, including authority and influence. Is it any wonder that this cat's greatest admirers seem to have been politicians, despots, royalty, religious leaders, shamans, actors, executives, sports figures, and the wealthy?

One can only speculate about the first encounters between people and jaguars. My guess is that early responses of *Homo sapiens* to *Panthera onca* were robust blends of esteem, wonder, and trepidation. Who would want to confront such a felid alone on a dark night? Yet over time humans developed the requisite weapons and skills that allowed them to keep jaguars at bay, and ultimately to capture and to kill them. Exactly how our ancestors accomplished this is unclear. Tragic and bloody consequences must have occurred over many years as hunters learned by trial and error to overcome the cat's might and cunning.

Over the past hundred centuries North America's climate has gradually warmed and dried along its lowland coasts and interior plateaus. In part as a consequence, the continent's population of game animals thinned. About seven thousand years ago, in many regions, the necessity for farming emerged. Here, as in other parts of the world, agriculture developed within a growing human population that had survived previously by harvesting wild game and native plants. Over millennia, accumulated survival skills and evolving social structures became the foundation for a series of Mesoamerican cultures impressive in their variety and complexity. Each stood on the shoulders of its antecedents, refining traditions and beliefs as need and circumstance dictated.

Beginning around 1500 B.C. the Olmec coalesced in southern Mexico as a distinct cultural group. It is not known how they referred to themselves;

the name we use today was assigned by the Aztec around A.D. 1300 and means "people of rubber," a reference to the Olmec use of latex tapped from trees. Facing the decline of naturally occurring foodstuffs, these people endured the slow process of domesticating squash, beans, chili peppers, tomatoes, and corn, as well as fine-tuning the firing of clay pottery. The Olmec also are credited as the first group in the New World to create a writing system, and they later explored mathematics and calendar-based timekeeping. From humble origins this civilization became dominant along the Gulf of Mexico from about 1200 to 400 B.C. The Olmec created city-state monarchies ruled by an elite class whose strategies for staying in power apparently included brutal human sacrifice. (Some of their descendants kept this tradition going until European arrival in the early sixteenth century.)

The Olmec spun an intricate web of spiritual traditions, celebrated at imposing religious complexes through construction and display of colossal stone heads, mud-plastered pyramids, and finely crafted ceramic objects. Retrieved artifacts from these centers provide clues about the relationship of these indigenous Mexicans to inhabitants of the natural world, particularly the jaguar. *Panthera onca* became a regal member of the Olmec pantheon of deified animals.

The most visible single collection of Olmec monuments is found today at the ceremonial site of La Venta, built near the Gulf of Mexico between 900 and 400 B.C. Visitors can see three large mosaics painstakingly laid out to form masks of ferocious-looking jaguars. Each assemblage, about fifteen by twenty feet in size, originally contained 485 blocks of beautifully carved serpentine stone. Nearby altars hold smaller jaguar icons. These and other carefully placed elements—including twenty-ton basalt statues believed to represent jaguar heads, and jade pendants that outline jaguar teeth—have led some archaeologists to speculate that the entire compound may have paid tribute to this one animal, perhaps in association with certain physical phenomena of nature.

"[The Olmec] were more than worshippers," writes Richard Perry in *The World of the Jaguar*. "They were jaguar psychotics—deforming their heads in imitation of the great cat's flattened skull and depicting the child-like or Mongolian faces of their images with feline mouths: the lower jaw brutally exaggerated."

What drove this preoccupation? Why did these people, in effect, worship jaguars?

One plausible explanation lies in an Olmec origin myth, which holds that the civilization's semidivine rulers were descended through acts of sexual intercourse—perhaps forced—between male jaguars and female humans. Such copulations are depicted graphically three known times in early Olmec art. The unions were thought to give rise, as described by the archaeologist Michael Coe in his influential book, *Mexico,* "to a race of were-jaguars, combining the lineaments of felines and men. These 'thunder-child' monsters are usually shown in Olmec art as somewhat infantile throughout life, with the puffy features of small, fat babies, snarling mouths, toothless gums or long, curved fangs and even claws." It was believed that were-jaguars subsequently took on human features and formed Olmec royal families, whose unchallenged hereditary power as "jaguar people" kept them in charge for centuries. Put simply, Olmec rulers may have convinced themselves that they were a crossbreed of cats and humans.

Physical evidence supporting the Olmec jaguar-royalty connection continues to surface. Some of the oldest known painted artworks in the New World are the roughly three thousand–year–old Juxtlahuaca Cave panels near the Gulf Coast, which show a bearded figure wearing spotted-fur leggings made from a jaguar pelt as well as a colorful headdress adorned with the plumes of a quetzal, the cloud forest bird prized for its iridescent feathers. Another Olmec cave painting—this one near Oxtotitlan, Guerrero—shows a jaguar and a woman in an apparent act of sexual intercourse.

Olmec jaguar images often are associated with mythic birds and godlike serpents, a triad that recurs consistently among other indigenous peoples as far north as Arizona's Apache. Deification suggests that the Olmec were among several early groups who connected jaguars directly to the gods or spirits they believed responsible for the sun, thunder, rain, night, and lightning.

By the time the Olmec civilization reached its zenith, around 500 B.C., jaguar imagery was common in such major population centers as Teoti-huacán, the sprawling ancient metropolis near present-day Mexico City. The feline's visage appears there in stone, ceramic, and metal figures, as well as sculpted monuments and heads. At least some Olmec artisans who crafted such objects were also priests and rulers. Through the creation of jaguar-themed art, it seems, an intrinsic connection between spiritual and political power became further enshrined. The details of how and why this link was made are unknown, yet potential clues still surface.

During 2006, for example, according to a report in *Discover* maga-zine, archaeologists working in the Veracruz region of Mexico found the skeleton of a 3500-year-old male Olmec who had his teeth filed down to stubs, apparently in order to wear a set of faux jaguar fangs. Presumably he did this in order to borrow the cat's status and abilities. James Chatters of AMEC Earth and Environmental, who examined the remains, believes that the filed-down teeth are evidence of the oldest known dental procedure in the Americas and of ceremonial activity in Mexico. The man's bones were buried in a painted rock shelter, his teeth ground to the gum line. The "fang dentures" coincide with a number of images of humans with jaguar teeth that show up in Olmec iconography.

Far to the southeast of Mexico, the people of Peru's Chavín de Huastar—a civilization contemporaneous with the Olmec and, at five thousand years, believed to be the oldest in South America—used a stylized jaguar motif in ceramic art similar to that of the Olmec. In fact, archaeologists specu-late the Chavín may have been visited by Olmec or their agents, which would have required overland trips of thousands of miles across Central and South America or journeys by boat from present-day Mexico to Peru. In support of this theory, researchers believe that Olmec influence is also visible among Mesoamerican pottery artifacts found in Guatemala, Costa Rica, and Panama. Like the Olmec, the Chavín are noted for their contri-butions to the early development of weaving, pottery, agriculture, religion, and architecture.

In 2006 it was reported that a Japanese researcher in northern Peru had excavated from the ruins of a lavish Chavín temple a large relief panel showing two jaguars. The cats' heads stand about five feet high and are believed to depict part of a myth of great significance, though the exact story and meaning have been lost to the ages. A salient feature of Chavín influence is the repeated representation of such stylized jaguars, hence the Chavín's frequent designation as a jaguar-worshiping cult. Before disappearing, they passed this proclivity to other cultural groups, including the Inca, who identified the jaguar closely with their creator god, Vivacocha, representing moon and night.

The Moche civilization, which flourished in northern Peru from about A.D. 100 to 800, also used the jaguar as a godlike symbol. Warriors adorned themselves with jaguar pelts or teeth as a way of enhancing courage and strength. The spotted cat also represented power and war through the guise of Moche ceramics and gold ornaments.

I learned in Ecuador that the Inca, the sophisticated civilization that ruled the northern Andes Mountains of South America from the early fifteenth century until the late 1600s, venerated the jaguar in cultlike fashion. The Inca believed, for example, that the physical contours of their capital, Cuzco, resembled a jaguar's body. Outside the city, huge rocks arranged like a cat's skull marked the Inca fortress of Sacsayhuaman.

As with the Maya and Aztec to the north, the most important members of the Inca empire hierarchy wore jaguar skins as symbols of their authority and used other parts of the cat's body for ceremonial and religious purposes. The animal's image adorned coats of arms and other official emblems. Masks and ceremonial cups, vases, and goblets rendered in the shape of jaguars were common, as were gold and silver objects. It is believed the jaguar symbolized "human present life" for the Inca, while other animals represented lives occurring only in the past and future.

According to some interpretations of Inca cosmology, the jaguar embodied the universal human ambition of living without enemies in this world and beyond. To that end, accomplished warriors associated themselves with jaguar imagery and jaguar-related adornments as a way of demonstrating dominance, beauty, power, and fearlessness. Although the Inca left no written language, the civilization's name for jaguar—*otorongo*—was assigned to its bravest soldiers. These included Otorongo Achachi, a celebrated hero who explored Amazon forests far from the Inca's highland stronghold.

Back in Mexico, the Olmec civilization eventually withered away. But it, too, seeded new cultures. One was the Zapotec, which flourished mainly in the present-day state of Oaxaca from around 500 B.C. to A.D. 1500. Zapotec culture also esteemed jaguars and linked the animals with the divine. One origin myth suggested that humans were derived from these felids and become them again after death. When the large burial crypt known as Tomb 7 was opened at the ancient capital of Monte Alban, near Ciudad Oaxaca, carved jaguar bones were found carefully arrayed, perhaps deposited to ensure everlasting protection for the departed ruler. Some believe that the entire site is essentially a citadel to Cosijo, the "jaguar god." The Zapotec, known for their elaborate religion and a form of glyphic writing, clashed eventually with the war-hungry Aztecs and lost power.

Many jaguar images are found among the stone monuments of the militaristic Toltec, who emerged in Mexico's highlands around A.D. 900 and held sway there for about two centuries. This civilization—possibly influenced by Maya migrants—was overseen briefly by a ruler named Four Jaguar. The Toltec are often credited with founding the first markedly urban culture of central Mexico, but, like the Zapotec, they were overrun by Aztec armies during the twelfth century.

In what is now Mexico's southern state of Chiapas, the short-lived Izapa civilization also embraced jaguar imagery, particularly in ceramic artwork and on monumental stone carvings that still remain in some locations. The abandoned city of Izapa, for instance, boasts imposing temple mounds and other features wherein late-period Olmec sculpture and early Maya paintings and glyphs appear to converge. From varied points of contact, the influence of Olmec art, religion, and governance filtered through the Izapa and other groups into (and from) the Maya civilization, which prevailed in southern Mexico and northern Mesoamerica shortly after the Olmec decline. As examples of this cross-fertilization, some images of early Maya priest-kings were carved on the obverse of Olmec jade pieces, and Maya royalty dressed in a style reminiscent of that adopted by the last Olmec rulers.

CHAPTER SEVEN

"Jaguars Possess the Power of God"

THE MAYA DEVELOPED ONE OF the most remarkable civilizations in human history. Over the course of centuries they built great cities, mastered astronomy, fashioned a sophisticated cosmology, devised a highly accurate calendar, and pursued large-scale agriculture in a demanding environment. They were talented artisans, accomplished builders, skillful politicians, fierce warriors, entrepreneurial traders, imaginative architects, articulate writers, and precise mathematicians. While Europe slept through its Dark Ages and suffered from the Black Plague, the flourishing Maya climbed to great heights—all without benefit of metal tools, beasts of burden, or the wheel.

Archaeologists divide the story of the Maya into five major chronological blocks stretching between 2000 B.C. and A.D. 1521. High tide was the Classic Period of A.D. 250 to 900. An estimated 3,500 known archaeological sites are connected with the Classic epoch and other Maya timelines, but only a small number have been examined thoroughly. Many of the conclusions we make about this civilization remain speculative and incomplete, yet it is abundantly clear that the Maya considered the jaguar an important symbol, supernatural protector, and sacred intermediary. In various contexts it was also an animal to be ritually sacrificed or "harvested" for its highly prized pelt, claws, teeth, and other body parts.

Contrary to a common present-day assumption, the Maya as a distinct society never died out. Millions of people of direct Maya descent retain age-old dialects, beliefs, and customs. Unfortunately, conquering Spaniards in the sixteenth and seventeenth centuries destroyed a great deal of material—including books, music, and religious rites—that probably explained the fine points of Maya cosmology. The knowledgeable ruling class was largely wiped out by warfare, starvation, and disease. Questions about where the original Maya came from, how they developed their civilization, and why their sophisticated social structure ultimately collapsed are among many still unanswered. By the time Spanish explorers came on the scene in the early 1500s, the Maya had left a vast area virtually empty in terms of human occupation, with once-majestic pyramids returning to the earthly forms from which they had sprung.

Speculation about the Maya has been clarified immeasurably since the 1970s, as carbon-14 dating technology resulted in the revision of age estimates for hundreds of artifacts and new scholarly interpretations of complex glyphic writings materialized. In addition, satellite imagery, radar, and infrared photography have allowed scientists to learn more about the land that sustained this civilization.

The Maya region had populations of astonishing size and became "one of the most densely populated areas in the pre-industrial world," according to the archaeologist Pat Culbert, emeritus professor at the University of Arizona. Some city-states in the Late Classic era comprised tens of thousands of dwellings, with densities as high as five hundred people per square mile. "The fields must have been crammed with workers, laboring from dawn till dusk," Culbert told *Archaeology* magazine in 2002. Maintaining social order over such a civilization had to be a daunting task, particularly as forests and wild game vanished in the effort to feed millions of people.

A consensus seems to have emerged among social scientists about political control among the Maya. It apparently rested in a powerful hierarchy of autocratic hereditary rulers—priest-kings, assisted by specialized shamans—whose strict governance melded all significant spiritual, military, political, and economic authority. A key component of the reigning cosmology was the jaguar, in whose precious skin Maya rulers wrapped themselves while seated on elevated thrones, feet tucked into moccasins made of jaguar leather. Sculpted limestone seats in Maya palaces were sometimes shaped like jaguars, covered with the cats' skins as a means of honoring both god and ruler. Jewelry and ornaments worn by priest-kings either reflected jaguar imagery or incorporated actual parts of jaguar bodies.

Over successive generations the jaguar was woven into a worshipful tapestry of art, religion, and legend. So strong was the feline's cultural aura that it continues to resonate in the region today. Modern Maya use the ancient word *balam*—sometimes spelled *balun* or *baalam*—to refer to the big spotted cat and often associate jaguars with place names, ceremonies, mythology, and authority.

"*Balam*—literally, 'tiger'—was also applied to a class of priests," wrote D. G. Brinton in his classic 1882 volume, *The Maya Chronicles.* The pioneering anthropologist noted that the boundaries between priests and rulers, jaguars and gods, were indistinct. Citing customs of the late nineteenth century, Brinton confirmed that *balam* "is still in use among the natives of Yucatán as the designation of the protective spirits of fields and towns." (I saw this myself one hundred years later in traditional Maya communities, where images of the jaguar often were displayed prominently in both public and private places.)

Scholars tell us that the jaguar dominates many Mesoamerican creation stories. According to the Mopan Maya of Belize, as recorded by J. Eric S. Thompson, the animal already existed when the "high god" called Itzamna began making the first humans out of clay. This effort failed when rain began to fall and the clay reverted to silt. Because he did not want the clever jaguar to watch his next attempt—involving the corn plant—Itzamna sent jaguar to fetch water from a river, carrying a hollow gourd punctured with small holes. Owing to his broad paws, the cat could not complete his assigned task. Frog hopped forward and convinced him first to plug the gourd's leaks with sticky mud and then let it dry. But this required so much manual dexterity and concentrated effort with its mittenlike paws that by the time jaguar returned, Itzamna already had made thirteen new men and a dozen weapons for their use. The creator deliberately endowed his corn-based humans with skill in weaponry, and man's intelligence was clear from the beginning. The envious jaguar was injured twice before conceding, with reluctance, that humans would be forever his masters.

Maya cosmology promotes the crucial importance of the balance believed to be inherent in all things. Without this intrinsic equilibrium, it was reasoned, life-sustaining systems would stop working and the natural world would suffer catastrophic collapse. In acknowledging bridges between the holy and the earthly, with Itzamna's prompting, humans took on the eternal burden of maintaining propitiatory ceremonies as a kind of religious theater. Such rites were accepted as a means of observing, celebrating, and retaining the essential order of nature. The Maya were convinced that without the ceremonial sacrifices they made, everyday miracles would either get uncontrollably out of adjustment or stop altogether.

In their seminal book *Maya Cosmos: Three Thousand Years on the Shaman's Path,* David Freidel, Linda Schele, and Joy Parker conclude that Hun-Nal-Le—the corn god that appears often in Classic Maya imagery—instigated the last of several origin events wrought by Itzamna, whom they describe as the greatest, most creative shaman of all. The archaeologists' study of ancient pottery led them to propose that the jaguar had an even more significant role than believed previously. On various painted vessels, the coauthors point out, the corn god is shown wearing a knapsack representing the cosmos: "He carries one [each] of three animals . . . a jaguar, a snake or lizard, and a monkey. These animals represented the three thrones that were set up in the first act of Creation." Summarizing the special bond between themselves and these animals, the Maya expressed a belief in coessences, a way of sharing spiritual as well as physical dimensions. Only creatures judged extraordinary were believed capable of becoming "spirit guides" for men and women in undertaking essential rituals.

Gene S. Stuart and George E. Stuart, in *Lost Kingdoms of the Maya,* refer to jaguar worship through the work of William L. Fash, past director of Copán's Acropolis Archaeological Project. Copán is a vast Classic Maya city tucked into the rolling green mountains of northern Honduras. Its finely crafted art and sophisticated monuments have prompted some scholars to dub it "the Athens of the Maya world." Evidence surfaced here, according to Fash, that "the jaguar was considered to be the intermediary between the world of the living and the world of the dead, and a protector and symbol for the Classic Maya royal houses." In 1988 excavations next to Altar Q at Copán appeared to have affirmed this theory when a masonry crypt yielded the remains of fifteen jaguars, buried in A.D. 775 by the ruler Yax Pac. Apparently, at the time of Yax Pac's burial, live jaguars were ritually sacrificed for each of the fifteen priest-kings who had preceded him in the royal dynasty.

In *Maya Cosmos*, Freidel and his colleagues describe Copán as a focal point for jaguar worship, "the place where the priest-kings conjured the gods and ancestors into this world through sacrifice. Flanking the stairs [of a large temple], which constituted the 'playing area' of [a] false ball court, are rampant jaguars, the spots of their pelts originally rendered with inset obsidian disks. They dance, just as the Hero Twins [who helped the first human survive] danced in victory over the Lords of Death." Known as the Hieroglyphic (or Jaguar) Stairway, this monumental structure continues to demonstrate in a dramatic way the protective role these cats played on behalf of Maya rulers.

The Classic Maya vaulted jaguar veneration to extraordinary heights. In its various guises, the cat is thought to have embodied several of the pantheon's most important deities, including those overseeing the sun, night, rain, and Xibalba—the surreal underworld where only the most holy and powerful men (and an occasional woman) could enjoy infinite afterlife. This accounts for the shattered pots and other burial artifacts I saw in Belize's Che Chem Ha Cave.

The jaguar god safeguarding Xibalba was a potent symbol of the presumed divine right of Maya royalty. This deity also ruled the Maya's "sunless sky," otherwise known as nighttime. According to priest-kings, the spots and rosettes decorating a jaguar's fur symbolized the splash of stars across the heavens, while allowing it to blend in with the shadows of trees. The cat's wide, perceptive eyes were said to gleam like the shining moon overhead.

A traditional Maya belief is that the jaguar god, as a ruler of darkness, is transformed daily into the fire-eyed sun god, a ruler of light. This occurs precisely at dawn each morning, when the sun emerges and begins to travel across the sky, once again becoming the jaguar god at dusk. A powerful deity is essential in order to take the sun safely through the forbidding night and draw it consistently beneath the darkened earth from west to east. Otherwise, the sun might never return. The jaguar, a fearsome predator that strides unchallenged during its nightly circuit, is a perfect choice for the job. Indeed, the Maya glyph for *night* shows the face of the sun with the ears of a jaguar, making it a kind of "night sun." In this metaphoric way, both the supernatural jaguar and the earthly priest-king defy the permanent death that afflicts less-exalted beings.

Jaguars were presumed to command a level of power to which the average human could never aspire. Only people of the highest status could expect to enjoy a few precious interludes of such mastery and strength. Under certain conditions a Maya shaman—accorded more spiritual capacities than a priest-king—was believed to transform himself, literally, into a jaguar. This ability to shift shape from human to cat and back to human again gave jaguars even more status among all Maya social classes.

Acts that seem outlandish by today's standards are perhaps better understood within their cultural context. The Maya were ruled by a cosmological order in which all things, including those not visible to the human eye, had specific energy and purpose. From such a perspective, it made sense for Maya leaders and warriors to wear the skin, teeth, and claws of jaguars during battle, since this is how they obtained special protection. For the same reason, Maya on the battlefield imitated the quick, lithe movements of the jaguar while roaring, hissing, and growling fiercely.

The Maya believed that the soul of every living thing resides in a creature's blood, the raw essence of life. Mortal sacrifice was apparently common during the Late Classic period, as ruling priest-kings and their duty-bound shamans tried in vain to keep their civilization from unraveling. Bloodletting was a means of appeasing or currying favor with the ruling spirits of nature. For such purposes, tongues and penises among the power elite sometimes were pierced with thorn-barbed strings and sharpened stingray spines, mixing blood and pain as gifts to the most important deities. On occasion jaguars would also have their blood drawn, or the animal would be sacrificed altogether. Without such devotion, the Maya believed, the full destructive wrath of displeased gods would be unleashed.

The ultimate sacrifice apparently extended to humans as well as animals. Some interpreters of Maya history say captured political enemies and prisoners of war were occasionally decapitated or tortured to death, or had their hearts ripped from their chests while still beating. In popular rubber-ball games, played on rectangular courts that can be walked upon today at many Maya sites, jaguar-derived deities presided over contests in which members of losing teams are said to have faced immediate torture and execution. As with similar propitiating rituals, such acts were meant to ensure that the gods would keep bringing rain and sun to the sky, fertility to the soil, and, by extension, life to humans through wild game they hunted and plants they grew.

Ritual bloodletting continues on a smaller scale among some traditional Maya of Mexico, Belize, and Guatemala. Instead of humans and jaguars, chickens and other livestock (along with strong sugarcane alcohol) are sacrificed in ceremonies generally closed to non-Maya. By invitation, I have been lucky enough to visit a few sacred sites where feathers, rum, and blood are found scattered among stone carvings and other objects representing specific gods. At locations near Guatemala's Chichicastenango, for example, I was allowed to photograph a shrine where shamans still routinely mix corn pollen and liquor with the fresh blood of small animals. Although I have not been among them, travelers in the Guatemala district of Chamula may witness during the annual Fiesta of San Sebastián the age-old ritual of the *balam tun*—"jaguar rock"—whereupon a Maya boy feigns sacrificial death on a large stone slab.

Researching five editions of my guidebook to Belize, a country that boasts a deep connection to all things Maya, I came to appreciate how well Alan Rabinowitz understood the long-standing relationship between jaguars and humans in the Cockscomb Basin where he worked. This was made manifest when the young scientist found overgrown abandoned ruins—later named Kuchil Balum (Place of the Jaguar)—and when he sought the incantations of a shaman named Miguel in order to draw a "tiger" to a baited trap. Rabinowitz recounts the Maya's ceremony in his memoir, *Jaguar:*

> The ritual lasted fifteen minutes. The room was filled with the pungent smell of copal smoke. Miguel stood, looking a bit drained.
>
> "De tiger cum close to you in tree days," he said to me quietly.
>
> "But will it come into my trap?"
>
> "I tell you what de [god] say—he say tiger cum close in tree days."
>
> . . . I paid Miguel and he asked for a drink, explaining that talking to the spirit world was a stressful task. I realized just how stressful it had been when during the next hour he finished the rest of my rum and started on my tequila.

Three days after Miguel performed this rite, Rabinowitz went to inspect the cage. The scientist was prepared for disappointment, since the cats had been ignoring his trap for months. But on this morning a jaguar was sitting quietly inside the enclosure, almost as if waiting patiently for its pursuer to show up.

For Miguel, this was no surprise. For at least twenty centuries, thousands of his ancestors had shared rich natural resources with jaguars along this fertile slice of Caribbean coast. Here Miguel's people have fished, hunted, farmed, worshiped, and built imposing structures of limestone block.

Ample evidence has surfaced throughout Belize of the jaguar's close connection with royal dynasties. During the 1980s, for example, a team of San Diego State University archaeologists excavating the ruins of Buenavista del Cayo unearthed an eighth-century crypt in which a young priest-king had been laid to rest wearing jaguar-skin clothing. Befitting his high status, the lord was buried wearing mittens made of actual jaguar paws.

Just across the border at northern Guatemala's most impressive ancient city, unknown to Europeans before 1848, lintel images displayed on Tikal's Temple I commemorate the "jaguar protector" known as Nu-Balam-Chakl. The cat's inscribed body looms threateningly over a priest-king, its gothic claws still reaching out to intimidate anyone who might dare rise against the supreme authority wielded by Nu-Balam-Chakl over Tikal's estimated 180,000 residents.

Nearly two centuries earlier, at Mexico's Palenque—an exquisite city-state sometimes compared in beauty and style to Italy's Renaissance-era Florence—explorers found carvings of life-size twin jaguars, their limestone backs used as thrones by a royal dynasty. Around the same time, carved heads of jaguars were documented about two hundred miles away at the Yucatán city of Uxmal on its House of the Dwarf temple.

Early Western explorers visiting Maya ruins were awestruck by what they saw; they refused to believe Native Americans were capable of making such fine things. Reflecting a condescending attitude that prevailed at the time of his 1846 visit, the U.S. adventurer John Lloyd Stephens insisted in his *Incidents of Travel in Central America, Chiapas, and the Yucatán* that "savages never reared these structures, savages never carved these stones." The monumental architecture was too finely wrought and sophisticated in design, Stephens believed, to have originated with the forebears of the indigenous people he encountered. Nonetheless, he was convinced that the Maya civilization was not a transplanted one. It developed and flourished, Stephens was certain, in the jungles and the highlands of Mesoamerica. But by whose hand? Stevens thought the constructions were the doing of some group long departed from the area, with no direct link to the Maya he met living on the fringes of such magnificent rubble.

It would be up to other Western outsiders, particularly the archaeologists of the late twentieth century, to acknowledge the ancient Maya's erudition and, with contributions from their subjects' descendants, to outline the complicated relationship the Classic Maya forged with their environment.

In northern Belize I once spent several days at the overgrown ruin of Lamanai, one of only a few Classic Maya city-states fully inhabited when Spanish colonials arrived in the sixteenth century. Lamanai's remote location, beside a swamp-encircled lake, discouraged foreigners and helped to keep traditional culture preserved. The Spanish adopted the site's original Mayan name, which translates as "submerged crocodile." Scores of the once-holy reptiles continue to slither through adjacent lagoons and marshes.

Walking beneath Lamanai's tall ficus trees, from whose boughs howler monkeys roared ominously, I was surprised to come upon the open-mouthed faces of two distinctly feline creatures rendered in limestone. The fearsome masks flank the northern stairway leading to the top of a sixty-foot-high temple that affords a stunning view of the entire area, now a patchwork of forest, swamp, and farmland. A thousand years earlier skilled craftsmen placed blocks and mortar in patterns clearly suggesting a jaguar's wide-set eyes, exposed fangs, protruding tongue, flat nose, thick brow, spotted ears, and upturned mouth. When first recorded by archaeologists in the 1970s, these molded features still had traces of the bright pigments that gave them lifelike colors.

"The carved jaguars stopped me in my tracks," I told the resident caretaker Nasario Ku, a Maya who worked closely with researchers who restored the masks in 2002. "They're haunting."

Ku, a diminutive man with slender hands and alert eyes, replied: "You're not alone, friend. Those tigers have protected that building for centuries. They must be doing a pretty good job, because they've withstood many attacks and are still very solid." Ku pointed out that his forebears had destroyed a sixteenth-century Catholic church and summarily executed its resident priests, souring the Spanish forever on Lamanai. Resident Maya later harassed British sugarcane farmers to the point that they, too, abandoned the site, which by then included a molasses mill colonials had constructed near the ruined mission during the 1800s. Today its rusted metal flywheel is entrapped by the octopus arms of a strangler fig.

Days later I walked through what little remains of Santa Rita, a Maya site engulfed by the modern, whitewashed town of Corozal, perched along an idyllic Caribbean inlet. Here, around 1900, the physician and amateur archaeologist Thomas Gann unearthed a catlike pottery figure with a human head in its jaws. It is one of only a handful of Maya references to possible jaguar attacks on humans. Small clay jaguars sitting upright were set around this particular object at the four cardinal compass points. Such placement echoes the traditional Maya practice of placing the carved head or effigy of a jaguar on each of the four sides of a village as a way to ensure its protection, a precautionary ritual common in the region well into the twentieth century.

Within the entirety of the Maya world, it is impossible to estimate how many temples and adornments have honored the jaguar in some way. One of the most well-preserved displays is visible at Chichén Itzá, the rambling Late Classic site on Mexico's Yucatán peninsula. Prominent throughout this city-state, which shows strong Toltec influences, are sculpted stone relief monuments depicting jaguars. At the Platform of the Jaguars and Eagles, a carving clearly shows the open mouth of a jaguar about to consume a human heart, while naked warriors torment the cat with flaming torches. The Temple of the Jaguars looms above an expansive central plaza and the Maya's largest known ball court. The temple gets its name from a cat-adorned carved frieze and a limestone sculpture of a jaguar, used as a throne by royalty. Originally painted a brilliant red, the chair shines with rosettes made of jade, eyes dotted by precious stones, and teeth taken from real jaguars. Even a casual visitor to Chichén Itzá will conclude that the people who built its monuments had an extraordinary relationship with the jaguar.

Farther south, the sacred link between jaguars and Maya is explicated at ruins scattered throughout Petén forests, and researchers continue to stumble upon new sites some two hundred years after formal investigation began. Vivid murals found in late 2005 at the San Bartolo ruin unveil with remarkable clarity the cat-themed Maya mythology of creation and kinship. The archaeologist Michael D. Coe has called the San Bartolo art "one of the greatest Maya discoveries of all time." In his 2006 autobiography, *Final Report: An Archaeologist Excavates His Past,* the retired Yale professor declared, "The great age of Maya archaeology is far from over. In fact, it's just beginning."

❦

Jaguar worship was hardly confined to the tropical lowlands where the spotted cat prowled most often. Around A.D. 1200, on highland Mexico's comparatively cool and arid central plateau, where wild jaguars were still being seen in 2009, an aggressive and power-hungry society began to dominate and absorb other cultural groups. Though they borrowed extensively from the traditions and cosmologies of antecedents, the Aztec people cobbled together a distinct civilization of their own. Like their predecessors, Aztec artists created figures in pottery and paintings that shared physical attributes of both humans and jaguars. They, too, gave the jaguar a prominent role in origin stories and considered the animal of holy origins.

According to an interpretation of the complex Aztec calendar advanced by the anthropologist George C. Vaillant, the first god of creation, Texcatlipoca, was transformed into the sun while fearless jaguars roamed Earth. The felids killed and ate the humans and giants who then populated our planet. Early in the process of creating Earth it seems that Texcatlipoca represented night and darkness, but he grew weary of living in the shadows. He turned himself into First Sun (a deity also called Smoky Mirror) and illuminated the world. This incurred the wrath of Quetzalcoatl, the feathered serpent god, who challenged Texcatlipoca to a duel. Quetzalcoatl succeeded in knocking his adversary into the sea, from which Texcatlipoca emerged as the Great Jaguar. As his fur dried, before he returned to the heavens, shiny black spots appeared on Texcatlipoca's coat. They remain visible in the constellation we now call the Great Bear (Ursa Major), but known to the Aztec as Great Jaguar. In this Aztec origin story the "earth world" is named Nahui Jaguar (Four Jaguar), one of a four-part cluster of primary elements that includes fire, water, and wind.

The jaguar also figures in an Aztec legend about the ancient city of

Teotihuacán, built twenty-five miles northeast of present-day Mexico City by an earlier culture and abandoned by it around A.D. 800. The Aztec believed their primary gods gathered in these ruins after the destruction of four previous worlds, each linked with a specific god and race of people. In fact, the name Teotihuacán translates loosely in Nahuatl, the Aztec language, as "birthplace of the gods."

Two of the surviving postdestruction gods—Nanahuatzin and Tecuciztecatl—decided to sacrifice themselves in a huge bonfire so that yet another world could be made. Jaguar and eagle, honoring the noble intention of these suicides, leaped into the flames as well. Nanahuatzin and Tecuciztecatl later emerged as the moon and sun, respectively, while jaguar and eagle arose Phoenix-like and were assigned leadership of the two great Aztec warrior clans. The animals' heroism is further immortalized as days of the month on the Aztec calendar, as well as in the soot-darkened feathers of the Mexican eagle and the charcoal-black rosettes of the jaguar.

A spiritual link with the spotted cat is reiterated through the Aztec royalty's establishment of an elite military order called the Jaguar Knights, sometimes referred to as the Tiger or Ocelot Knights. (The discrepancy in names derives from the original Aztec word for jaguar, *ocelotl*. Over time, *ocelot* became the preferred term for describing the small spotted cat overlapping much of the jaguar's range.)

Warriors of the Jaguar Knights, identified by the pelts and masks they wore in combat, were allowed to take part in Aztec war councils, clan cabinets, and regal summit meetings, as well as special dances and ceremonies. They paid neither taxes nor tribute, and the finest among them were rewarded generously with land grants and war booty. Jaguar Knights participated in human sacrifices and cannibalistic feasts in which victims were often captured enemies, killed in slow, gruesome rituals. At the Museum of Anthropology in Mexico City, an Aztec receptacle on display is shaped like a jaguar and made for the discrete purpose of holding the hearts of human sacrifice victims.

By the time the Spanish invaded Mexico's central highlands in the sixteenth century, another elite warrior society called the Eagle Knights had emerged, along with a smaller group known as the Arrow Knights. Some experts believe that the two larger societies represented day (eagles) and night (jaguars). This symbolism may derive directly from the Maya, for whom jaguars were emissaries of darkness and protectors of the under-world. The Aztec warrior triumvirate, however strong, could not prevent disintegration of the empire at the hands of Spanish conquistadors, who began a swift takeover following the arrival of Hernán Cortés in 1521.

In dispatches to Spain, Cortés noted the great respect accorded jaguars by Montezuma, the Aztec ruler. Spanish scribes reported that the supreme leader received Cortés on a throne draped with a jaguar pelt and cushioned by eagle feathers, a seating arrangement much like that adopted centuries earlier by Maya priest-kings. A tour of Montezuma's private zoo by Cortés revealed handmade bronze cages holding jaguars and other wild cats, said to have been fed the bodies of sacrificed Aztec enemies whose still-beating hearts had been removed during ritualized executions.

In various murals, sculptures, and monuments, the Aztec jaguar continues to be displayed with pride throughout Mexico. The cat wears a feathered headdress as it stands alongside Quetzalcoatl, the mighty "plumed serpent" deity of the Olmec, which remains a defining cultural icon. Quetzalcoatl adorns the country's flag as well as thousands of calendars and posters hanging on walls inside Mexican homes, schools, and offices.

In addition to the nation's mixed-race majority, some of Mexico's extant indigenous groups continue to revere jaguars. Notable are the Tarascan and Huasteca tribes, ensconced in the north-central mountains. Elsewhere, folk traditions honoring jaguars mingle with Catholicism. The ongoing jaguar dances of some rural villages are believed to have origins in those performed by the Aztec centuries ago. In Suchiapa, for example, the Catholic Church's Corpus Christi festival includes dozens of teenage boys wearing jaguar masks. Their job is to run to the front of the annual street procession, securing intersections so that the parade moves through town smoothly. The shouts of these wanna-be jaguars—flexing their power just as Aztec warriors did in the sixteenth century—can be heard for blocks around.

The highly spiritual Huichol people of Nayarit and Jalisco continue to make and use bead-and-beeswax jaguar heads and masks as spiritual totems associated with the phenomena of rain and masculine power. Huichol traders fan out across much of Mexico and parts of the United States to sell these creations to tourists and collectors as decorative objects. I once asked an artisan hawking such masks in the Pacific Coast village of Sayulita, Nayarit, to explain why his tribe ascribed so much authority to jaguars. The old fellow smiled, then lifted his hands and held them apart in the international gesture of being overwhelmed. His dark face crinkled by a lifetime of smiles, laughter, and sunshine, the man leaned forward and whispered: "Por un hombre solo, eso es imposible, señor, porque los tigres tienen el poder del dío"—By one man alone, that is impossible, sir, because jaguars possess the power of God.

CHAPTER EIGHT

"Blood of the Valiant"

RESPECT MAY BE LONG-STANDING, but so is our enduring compulsion to hunt, capture, possess, and kill jaguars. For millennia, people have sought to eliminate, or at least control, these carnivores. The biologist Ronald Nowak, citing research by Alan Rabinowitz and others, wrote that when combined with other human-related factors, "direct persecution and declining prey have rendered [jaguars] among the most imperiled big cats." Rabinowitz himself concluded: "Jaguars and people can coexist in this world, but we all must work a lot harder at making this happen."

Because of the cat's vaunted cultural roles and the sheer magnificence of its pelt, the jaguar has been a perennial object of desire. It has been assigned great value in trade, fashion, and ritual for thousands of years. Mesoamerica's Olmec, Aztec, and Maya civilizations coveted jaguar pelts as key ceremonial objects, and in more recent times pelts have been enormously popular in the making not only of coats and purses but also of rugs, wall hangings, cowboy chaps, mittens, quivers, shoes, and other status-conferring decorations. In many cultures, the bones, teeth, and claws of the cat are equally desirable. And, of course, a "trophy mount" of a jaguar's snarling head or outstretched body has been among the highest goals of many big-game hunters.

One practice that has ended, perhaps forever, is the wholesale slaughter of jaguars in order to provide products for the luxury clothing and accessory market. Through much of the nineteenth and twentieth centuries, thousands of jaguars were harvested each year by the high-end garment industry. Hats, gloves, coats, and handbags made of jaguar pelts were the height of style in chic circles from the 1920s until the early 1970s. Aggressive harvesting spurred by this trade may have contributed to the jaguar's early extirpation from El Salvador, Chile, and Uruguay and its near eradication from several other countries. The peak of interest in spotted cat furs was probably reached in 1968, when the number of jaguar pelts entering the United States reached 13,516. That year some 50,000 leopard skins also were shipped worldwide, including about 10,000 to the United States. Reflective of their cachet in this era was America's fashionable first lady, Jackie Kennedy, who proudly wore a custom-tailored leopard-skin coat.

In 1969 an estimated 9,831 jaguar skins valued at $1.5 million were imported legally into the United States, at a typical price paid to hunters of $80 to $500 per skin. Designer coats sold for up to $20,000 at the time and fueled an industry generating an estimated $30 million annually. Some scientists believe that jaguars—along with their smaller cousins, ocelots and margays—might have been wiped out in the wild had such commerce continued unabated.

By then cat-skin harvesting had gone on with great enthusiasm and nearly no controls for more than a century. Beginning in the mid-1800s, at least four thousand jaguars were reported killed for export across South America each year. These numbers grew larger annually, and by 1900 an estimated two thousand pelts were shipped to Europe from Buenos Aires alone. Hunting pressures grew enormously in Amazonia. During 1966, according to the Peruvian government, skins traded officially at a single upper Amazon port, Iquitos, Peru, numbered at least fifteen thousand ocelots, four thousand margays, and nine hundred jaguars. Most pelts originated in Brazil, although other countries in South and Central America, along with Mexico, collectively contributed thousands.

Commercial jaguar trade has been illegal between most countries since July 1975, when the cats came under strict protection from Appendix A of the Convention on International Trade in Endangered Species, an agreement commonly referred to by its acronym: CITES. The accord's primary goal is to ensure that international commerce does not threaten the survival of wild plants and animals as species. The convention draws on current scientific consensus to determine the degree to which any among thousands of species merits protection. Although CITES is binding on its signatories, it depends on legislation and enforcement in each participating nation. Some countries in the jaguar's range seem far from eager to crack down on those who sell the cat's body parts on the global market.

In 2009 a prime illegally traded jaguar pelt was said to fetch fifteen thousand dollars, with custom coats costing at least 50 percent more. Despite well-organized campaigns against the industry's use of wild felids, the demand for "exotic cat" fur coats enjoyed a gradual comeback starting in the 1990s as these status-conferring luxury items regained their appeal. Most are now "faux fur" made of synthetic material, but some cat species—not including jaguars—are raised commercially for their skins. Any authentic jaguar furs currently sold legally are authenticated as "vintage" (produced before 1975).

An increasing number of nonprofit groups and government agencies promote jaguar conservation, research, and education, which also is believed to help keep trafficking in pelts low compared with the pre-CITES era. Between 1990 and 1994 only sixteen illegal skins were reported in U.S. commerce, and between 2000 and 2005 federal charges were brought against fewer than a dozen companies and individuals for the illegal sale of jaguar coats or artifacts. Even so, reliable reports and my own interviews confirm the sale of jaguar pelts by hunters in much of Latin America. While traveling in the region I have regularly received first- or secondhand reports of skins being bought and sold on the black market.

Jaguar-related commerce in the United States is overseen by the Fish and Wildlife Service (USFWS), an agency of the Department of the Interior. Importation of live specimens by zoos is subject to careful review by USFWS, while sale of nearly all other jaguar material is now banned. The agency has prosecuted sport hunters who tried to circumvent the law by mislabeling jaguar trophies as nonrestricted animals in order to sneak them past U.S. Customs agents. In one instance, a hunter brought from Venezuela a jaguar skin appraised at eleven thousand dollars by packaging it as a goat hide valued at a mere sixty dollars.

Internationally, illegal wildlife trade is a multibillion-dollar business, said to be third in dollar value worldwide to arms and drugs sales. Traffickers know that enforcement efforts in much of the world are meager. A jaguar stronghold like Colombia, for example, is said to employ only one hundred wildlife inspectors spread through a nation with a border that extends almost twelve hundred miles and an area bigger than Texas and California combined. As a result, according to informed sources, collusion between ranchers and pelt dealers in Colombia, as in several other South American countries, is common.

An escalating problem that reportedly affects all large felids is use of their penises, blood, fat, teeth, bones, and other body parts in Asian medicine or in concoctions that claim to be aphrodisiacs. In response to the increased scarcity of tigers—whose bones, teeth, claws, and penises are particularly prized in Southeast Asia—poachers are said to be turning to jaguars and other big cats in order to pass off harvested body parts as acceptable tiger substitutes. Assertions about the trend's significance surfaced several times as I traveled through the Americas, particularly in the growing number of areas with links to China through trade and immigrant communities.

Meanwhile, the consumption of jaguar meat appears to remain a rare occurrence. Archaeological sites show few signs of the cats being cooked or consumed by ancient humans, and only rarely do reports surface of people eating jaguars today. Exceptions include some tribal people of the Amazon who are known to eat jaguar tongues as a symbolic source of strength and to collect the cat's blood for ceremonial use.

In F. Bruce Lamb's *Wizard of the Upper Amazon,* a Peruvian who lived for many years among indigenous tribes describes an instance in which a jaguar was killed in order to harvest its blood for a kind of christening. Manuel Córdova-Rios told Lamb that around 1910 he accompanied rain forest hunters in a successful blood collection from a jaguar's jugular vein. The liquid was used to commemorate the birth of an infant. "After the cat was killed," Córdova-Rios recalled, "the jaguar chant was begun and the chief anointed the newborn boy with the blood of the valiant jaguar. The chief gave him the name Iria (Leader) and predicted hunting prowess and leadership ability for him in the future."

Under terms of the Captive Wildlife Safety Act of 2003, live jaguars cannot be sold in the United States to private individuals or transported across state lines as pets. While this federal law does not preclude in-state breeding and sale of jaguars, in 2009 some twenty states had outright bans on keeping large exotic cats as pets. Despite such legislation, the Humane Society of the United States estimated in 2009 that 10,000 to 15,000 big cats were privately owned across the county, including as many as 15,000 tigers. Florida alone was home to a reported 1,455 privately owned tigers at a time when only 5,000 to 7,000 tigers were believed to remain in the wild.

Jaguars, by comparison, are rarely kept as pets in the United States and seldom held past the cub stage elsewhere. However, the practice does happen. In August 2007, for example, a three-month-old cub was offered for sale at twelve thousand dollars by a pet store in Juárez, Mexico, immediately across the border from El Paso, Texas. According to press reports, local police arrested the shopkeeper and confiscated the jaguar.

The *hunting* of jaguars—legal and otherwise—is another story.

Some who know him were not surprised when John L. Klump became the first person in modern history to be suspected of killing a wild jaguar in the United States. In December 1986 the Bowie, Arizona, rancher reportedly tracked a healthy adult male with trained hounds through the uninhabited Dos Cabezas Mountains before shooting the exhausted animal. Second-hand reports held that Klump paraded his kill through the nearby ranching town of Willcox on the hood of his pickup and bought celebratory drinks at a local bar. These were daring moves, even in a hunter-friendly corner of the Southwest.

"John never did follow the rules, even in high school," said a former classmate, who grew up near Willcox and talked about Klump only after

I promised not to reveal his name. "I've lived in Cochise County all my life," the middle-aged man explained, with a nervous downward shake of his head. "I don't need to get on the wrong side of the Klumps."

Klump family members control grazing on nearly 300,000 acres of Arizona and New Mexico land, making theirs a dominant ranching dynasty in the border region. Driving through Cochise County on Interstate 10, travelers may pass a Klump-sponsored billboard commanding, in big block letters, "EAT BEEF."

Perhaps owing to his clan's considerable influence—and despite circumstantial evidence against him—John L. Klump was not arrested on wildlife-related charges for nearly seven years. Even after his apprehension, the 1986 jaguar shooting never figured directly in the rancher's prosecution. This was despite a posted reward of more than four thousand dollars and claims by a number of area residents that they had seen either photographs of the cat in question or its actual remains. Indeed, a widely published Arizona Game and Fish Department picture shows a man identified as Klump standing next to a lifeless jaguar hanging upside down beside a tracking hound.

On St. Patrick's Day 1993, undercover agents arrested Klump and his fellow hunter Timm J. Haas in a motel room near Albuquerque's main airport. The pair was nabbed in a sting operation for allegedly accepting thirteen thousand dollars in the sale of a mounted jaguar, jaguar rug, mounted ocelot, and ocelot hide, presumably brought into New Mexico from Arizona. The pair was charged with conspiracy and violation of the Lacey Act, which prohibits trafficking in such protected species across state lines. The two men also were accused of illegally killing black bears, javelinas, and desert bighorn sheep.

Klump and Haas initially pleaded guilty but withdrew their pleas and delayed further prosecution with 111 legal motions and two venue changes. On November 25, 1998, a Tucson jury found the men guilty, and more than six years later the defendants were sentenced to pay a fine and give up seized property. They served no prison time.

The government cases against Klump and Haas were stymied by the fact that jaguars at the time of the killing were not covered by the federal Endangered Species Act, created in 1973 to protect U.S. plants and animals determined to be at greatest risk of extinction. Bureaucrats explained that they had followed the conventional wisdom that *Panthera onca* had been extirpated.

"We had protected the jaguar south of the border but failed to protect them in the United States," recalled the retired FWS wildlife biologist Ron Thompson in an interview published in the Southwest Jaguars blog. "When I saw the mounted [Dos Cabezas] animal and the photos [of the dead cat], I was amazed at how an animal had traveled so far and at how the mighty Endangered Species Act could not protect it."

A further complication was the presumed killer's adamant refusal to reveal the whereabouts of the jaguar's hide. "It was," Klump is said to have told authorities, "the greatest trophy of my life." Of the two *Panthera onca* trophies retrieved, one was identified from its rosette pattern as the Arizona cat, the other as a jaguar taken in Mexico. (Since this case was prosecuted, the official list of endangered species in the United States has been amended to include jaguars, although federal agencies have been criticized for not establishing stronger protections for the animal.)

Some Klump family members espouse antiauthoritarian attitudes, once common in the Southwest, that are increasingly anachronistic even among staunchly conservative ranchers. In a case that drew national attention, Wally Klump—John's uncle—spent more than a year in an Arizona state prison after a 2003 contempt conviction. His crime? Refusing to comply with a judge's order demanding removal of cattle from a Bureau of Land Management grazing allotment. Seventy years old at the time of incarceration, Wally Klump was quoted by the *New York Times* as telling BLM employees: "You take these cows [of mine away], I'll kill you as mandated by the Second Amendment." According to news reports, this threat was repeated to a U.S. attorney prosecuting Klump's case. A heart condition eventually led to a change in plea, compliance with the BLM order, and the prisoner's release.

The Klump cases illustrate the kind of Old West standoffs that still occur over U.S. management of public land and wildlife. Some Southwest ranchers worry that carnivorous predators designated as "endangered" threaten their livestock operations, which frequently depend on renewable permits for grazing rights. (Livestock grazing leases are administered by the federal Bureau of Land Management or the U.S. Forest Service, which charge ranchers varying fees for use of lands the agencies control.) The government has the power to restrict access to and regulate activities on federal public lands it designates as "critical habitat" for endangered species. The fear among some ranchers is that cattle could be ordered off some grazing allotments in order to protect such land for jaguars.

So far, with respect to the jaguar, such fears have been unfounded. The Fish and Wildlife Service in July 2006 opposed such a move, stating that "based on a thorough review of all available data . . . there are no physical and biological features in the United States that meet the definition of critical habitat (for *Panthera onca*)." Such a designation would not be prudent, the agency said, because "U.S. habitat is believed to be marginal . . . and represents less than one percent of the species' current range." It concluded that "preservation and recovery of the jaguar depends almost entirely on conservation efforts in Mexico and Central and South America."

The Fish and Wildlife Service was promptly sued by the nonprofit Center for Biological Diversity, which insisted that designating critical habitat for jaguars would help the species regain its foothold north of the border. The FWS decision was "based on purposefully inadequate information and erroneous logic," declared Michael Robinson, a Center spokesman. "The longer the government stalls, the harder it will be to recover the jaguar." But in 2008 the FWS reaffirmed its earlier conclusions in a matter involving construction of border barriers, reiterating that conservation of the species should focus on activities south of the international boundary. The U.S. Supreme Court later declined to consider a legal challenge to the Department of Homeland Security policy of waiving existing environmental rules in building borderline walls that would potentially interfere with wildlife.

❦

For some border-area residents, the return of the jaguar is unwelcome. If a sealed border with Mexico keeps them from coming north, so be it. "They are damn varmints," one rancher told me. "Pure and simple."

Invoking the word *varmint*—defined as "an objectionable or undesirable animal, usually predatory"—hearkens to the early twentieth century, when government agencies classified jaguars as unwanted livestock predators and said that the creatures could be killed by anyone for any reason throughout the year. In Arizona this "nuisance wildlife" policy mandated an open season until 1969, when it was presumed jaguars had been fully eliminated from the state.

Like the wolf, coyote, bobcat, bear, and mountain lion, jaguars were long considered threats to cattle, sheep, and other livestock. In the early 1900s government agents typically paid five dollars for each jaguar kill, as reported in *Borderland Jaguars,* which details the history of the species along the U.S.-Mexico frontier. The authors David E. Brown and Carlos López González cite county ledgers and game reports in the Southwest dating from 1913 to 1920, which occasionally listed jaguars as "leopards" and "tigers."

Federal authorities, through the U.S. Bureau of Biological Survey, also were charged with exterminating predators. The agency's 1930 Arizona policy manual declared: "All Lobo wolves and jaguars will be taken as fast as they enter this State from Mexico and New Mexico, as 100 percent of them live on livestock and game."

No human-caused killing of a wild *Panthera onca* specimen has been reported in the United States since the 1986 Klump incident, but south of the border jaguars are routinely taken. In ranch country throughout its range, the jaguar continues to be widely regarded as a varmint.

Figure 8.1. Caroline Brown, smiling daughter of the Tucson saloonkeeper C. O. Brown, sits on the corpse of a big male jaguar killed in Arizona's Rincon Mountains in 1902. Courtesy Arizona Historical Society/Tucson; image #51506

"There are no finer people than *rancheros* and *vaqueros* of the Sierra Madre," says the biologist Thompson, who has assisted both hunters and researchers in the backcountry and northern Mexico, "but some [ranchers and cowboys] get up every morning wondering what has eaten their livestock during the night, for it may mean the difference of what they [themselves will] eat that night."

Even some conservation-minded activists and researchers concede that there are practical rationales for killing jaguars and other big cats on occasion. Controlled, legalized hunting may be deemed an appropriate management tool when too many people, wild carnivores, and domestic animals compete for the same resources. But government officials, ranchers, farmers, and researchers throughout the Americas also have collaborated in devising nonlethal techniques for mitigating jaguar depredation. The solution to the depredation problem, they believe, often lies in cooperation rather than angry confrontation. Some such strategies, ranging from modified livestock management to compensation for depredation losses, are outlined later in this book.

The sixteenth-century arrival of Europeans, with their formidable guns and special tracking dogs, expanded jaguar hunting as a popular sport. For instance, during the early Spanish colonial era it was reported that some two thousand jaguars were killed annually along the Río de la Plata, a river shared by Uruguay and Argentina. Another two thousand were taken each year in nearby Paraguay. Cats often were shot by prominent politicians, ranchers, and businessmen during elaborate hunting expeditions. In part as a result, by the late nineteenth century the region's jaguars were found routinely only in the least-populated forests, wetlands, and plains of interior Brazil, Paraguay, and Bolivia.

Jaguar hunts reached the height of their popularity during the first seventy years of the twentieth century, when professional guides earned good incomes and great prestige by helping clients track and kill the cats. The former president Theodore Roosevelt and the matinee idol Clark Gable were among those who paid top-dollar for elaborate escorted safaris that sometimes boasted portable refrigerators, crystal glassware, and linen tablecloths. While Gable's 1938 trip to Sonora failed to yield a single jaguar, Teddy had better luck in South America.

Roosevelt's 1914 expedition up an Amazon tributary—in part under the pretext of collecting specimens of tropical flora and fauna for the American Museum of Natural History in New York—inspired him to write a best seller entitled *Through the Brazilian Wilderness*. The book inflamed a growing passion among the wealthy for jaguar hunts escorted by professional guides. A famous still photo (and short movie) taken during his trip to the Pantanal region of southern Brazil documents the hunter-conservationist's shooting of an adult male. Roosevelt, wearing a pith helmet, is shown holding the head of a dead cat in one hand and a rifle in the other. The ex-president's adventures are recounted in characteristically florid prose in his story "A Jaguar-Hunt on the Taquary."

The jaguar, Roosevelt wrote, is "the king of South American game, ranking on an equality with the noblest beasts of the chase of North America, second only to the huge and fierce creatures which stand at the head of the big game of Africa and Asia. The great spotted creatures are very beautiful. Like all cats they are easily killed with a pack of hounds, but they are very difficult to come upon otherwise. They will charge men and sometimes become man-eaters." Despite his notoriety as an amateur naturalist, some of Roosevelt's assertions—notably about the jaguar as a "man-eater"—were inaccurate. Unfortunately, the popularity of his writing perpetuated an unjustified fear of jaguars that persists widely today.

On the other hand, Roosevelt is said to have been the first to observe that jaguar depredation on livestock in Brazil was more common on ranches with a scarcity of wild prey, while occurring less often in places where such prey was abundant. This notion was supported by several scientific studies undertaken more than seventy years later.

One of the most successful guides of the "jaguar safari" era was Dale D. Lee, the most celebrated member of a family of big-game hunters and wildlife trackers. Born in 1908, Lee spent most of his childhood on the east slope of Arizona's Chiricahua Mountains in the tiny village of Paradise, where his mother ran a hotel. The isolated peaks looming overhead, the last hideout of the Apache warrior Geronimo, teemed with wildlife, including some species found nowhere else. A crack shot, Lee took his first deer at age thirteen and within a few years became a full-time bounty hunter, paid by the State of New Mexico to track and kill mountain lions.

As adults, several of the Lee brothers became professional hunting guides, traveling from Alaska to the Amazon with their premium

Figure 8.2. Theodore Roosevelt poses beside a jaguar the former U.S. president shot and killed during a 1914 expedition to the Pantanal district along Brazil's border with Paraguay. Note tracking hounds. Courtesy American Museum of Natural History; image #104904

hounds in search of game animals and adventure. In *The Greatest Guide,* a 1981 series of vivid recollections compiled by his fellow hunter Robert McCurdy, Dale Lee claimed that he and his family collectively had dispatched more than 1,000 bears, at least 1,000 mountain lions, and 124 jaguars, along with dozens of bobcats, ocelots, and jaguarundis (a dog-sized tropical cat). A skilled and relentless tracker, Lee, along with his canines, would spend weeks pursuing wily predators through the roughest terrain imaginable.

The Lees raised their prized dogs on remote ranches in southeastern Arizona, where the scents of captive adult bears and lions, seized as cubs and kittens, were used to teach the hounds how to track and bring animals to bay. Apparently no jaguar was ever held for this purpose.

On many of the Lees' hunts a "caller" made of a dried hollow squash and leather straps was used to imitate a jaguar's distinct vocalizations. As described in *Borderland Jaguars,* a Lee brothers promotional flier from 1950 outlined the caller's intended result: "The jaguar's roaring in answer while approaching ever nearer is guaranteed to give you a man-sized thrill no matter how much big-game hunting you have done!"

In the northern Belize village of Caledonia, I once spent an afternoon with a traditional Creole medicine man, locally called a "snake doctor," who showed me a caller fashioned from a hollowed gourd, the widest end of which was sliced open and covered by stretched buckskin. A taut piece of sinew was stretched from one end of the calabash to the other and "played" by rubbing fingers along its length. It emitted a guttural sound akin to the deep-throated roar of a mature jaguar. A cat is drawn by such a caller into what it anticipates will be an encounter with another of its species, only to be surprised and chased by fast-pursuing hounds and armed men. Hunters are usually mounted on mules or horses, providing an advantage in strength and stamina. Live bait or fresh carrion also may be staked to attract a jaguar, which is then shot by a sportsman hidden in a blind or on a camouflaged platform. The usual weapon is a high-powered rifle, though a bow and arrow might be employed on occasion. In the Amazon, spears and machetes also have been used for the coup de grâce.

At age twenty-seven, Dale Lee went to the Sierra Madre of Sonora for his first jaguar hunt. Accompanied by a pack of hounds and an older brother, Clell, Dale drove a female cat into a cave near the Río Granados, where a breeding population of jaguars still exists. Each brother took shots at the cornered jaguar, a trophy they subsequently shared. The men later returned to the region for several successful hunts.

In later years Dale Lee shot jaguars as heavy as 250 to 275 pounds in South America. During his lifetime, the marksman guided jaguar expeditions in Colombia, Venezuela, Bolivia, Argentina, Brazil, Belize, Nicaragua, and Costa Rica, as well as Mexico. In all cases, Lee told McCurdy, the escorted sportsmen wanted to bring back trophies, either for mounting by a taxidermist or as skins to be displayed on their walls. The cost for this

Figure 8.3. The renowned big-game hunter Dale Lee, who guided clients through-out Latin America on hunts for exotic cats, displays jaguar hides taken during 1935 in Sonora's Sierra Madre Occidental. Courtesy Arizona Historical Society/Tucson; image #2113

privilege in the late 1940s and early 1950s was two thousand dollars for a thirteen-day expedition outfitting two hunters.

In 1964 a Dale Lee–sponsored jaguar hunt in British Honduras (now Belize) was aborted when government officials could not be placated, and the guide decided to stop hunting the species altogether. The following year he took a job as game ranger for the seventy-three thousand–acre Fort Huachuca military reserve in southern Arizona. Lee went on his last mountain lion hunt in 1984 and died in 1988 at age seventy-nine.

Hunters such as Dale Lee were products of their time, place, and culture. They grew up in ranch country during an era in which jaguars and other predators were shot on sight. "Most hunters of his day viewed our nation's wildlife as a never-ending population that could be made use of for profit," Jack Childs, a friend of Dale Lee's and fellow mountain lion hunter, told me. "Thankfully, our wildlife is managed under more realistic guidelines today." Childs believes that if Lee—whom he recalled as "an exceptional woodsman and hunter"—were alive today, "he would be a champion for jaguar conservation in the United States. We could have learned a lot about jaguar biology from Dale."

Childs conceded that while "some would disdain the way [Dale Lee] earned his living, he did it with pride and stayed within the bounds of the laws of the day. He truly rose to the top of his profession."

The pressure of sport hunting on large predators has declined to an extent in the United States. Some observers believe that this is because of a wider understanding of the key role such animals play in controlling the number, behavior, and health of deer, elk, and other prey mammals. Recreational hunting in general has declined in popularity as the population has become more urbanized and younger people have shown less enthusiasm for the sport. At the same time, government management of wildlife has become more closely aligned with scientific research while staying responsive to the needs of ranchers, hunters, bicyclists, hikers, and birders. In some cases, as with the mountain lions of Colorado, large predators have recovered (and moved into human-occupied landscapes) to such a degree that legal hunts are deemed both appropriate and desirable in order to thin numbers.

Unsanctioned killing of jaguars south of the U.S. border has slowed but not stopped. Continued poaching reflects a constellation of factors: strong traditions, misinformation, lax law enforcement, government corruption, bribery, sale of pelts, and loopholes in hunting regulations. People in jaguar country also kill big cats because they fear for the physical safety of themselves, as well as of their families, neighbors, pets, and livestock. Whether such fears are warranted is a matter of debate, and the situation differs from one country to another.

Mexico is a useful example. Before 1966 Mexican authorities classified *Panthera onca* as a predatory game species with a prescribed open season and bag limit. Special hunting permits were issued between 1966 and 1987, after which the federal government banned all jaguar sport-hunting and declared the animal to be in danger of extirpation. In spite of this measure,

Figure 8.4. The Arizona guide Dale Lee, left, poses with clients, assistants, and trophies after a successful jaguar hunt in Mexico, date unknown. Courtesy Arizona Historical Society/Tucson; image #49482

and the country's post-1992 compliance with CITES, Mexico maintains a separate law that allows landowners to kill any wild animal—including a jaguar—deemed a threat to personal property.

Documentation of such depredation varies. If a Mexican rancher declares that a jaguar has killed his calf or lamb, he may shoot it with relative impunity even when no proof can be shown that a cat was responsible. Wildlife agents in Mexico have little enforcement authority and rarely carry weapons to back up their citations. "Game wardens wield almost no power," a former Midwest state game officer who has studied jaguars on Mexican ranches told me. As a result, large predators enjoy widely varied degrees of protection in the countryside, including the area south of Arizona, from which most experts believe that state's minuscule population of jaguars is derived. (Some scientists feel that a few U.S. jaguars now may be well established north of the border.)

Unauthorized hunting is an ongoing problem in virtually every country where wild jaguars exist, particularly Venezuela, Bolivia, Paraguay, and Brazil. These latter nations have strong livestock industries that leverage considerable political and economic clout. If a South American stockman wants to kill a jaguar on the pretext that it poses a threat to his livelihood, neither neighbors nor authorities are likely to intervene.

I was told by several researchers working in Latin America that the only significant shift since the shutdown of international jaguar commerce in the mid-1970s may be a greater reluctance among jaguar killers to discuss their harvesting practices openly. But these researchers also see more willingness among ranchers to cooperate with scientists and government officials who are trying to study and protect big cats. Some ranchers in jaguar country have developed ecotourism lodges and guide services that promote wildlife sightings on or near their properties. Foreign tourists, particularly from the United States, Canada, and Europe, show a growing appetite for nature-oriented travel that includes the chance to see a jaguar on its home turf. These visitors may stay on a working cattle ranch and be led by working cowboys to locations where they can spot wildlife easily. In some instances, field biologists are persuaded to participate in the tours as a means of promoting conservation and soliciting donations.

A resolution of predator-related conflicts is being sought by an increasing number of conservationists and government officials. Some scientists and policymakers have concluded that ranching in some areas may be a relative boon to jaguar conservation inasmuch as it keeps in good condition large parcels of habitat that otherwise might be developed for less compatible uses. If killing of jaguars can be stopped—or kept at a manageable level—this may the most viable approach in places where protected landscapes need to earn their keep.

But the status quo reportedly has not changed in much of Latin America, and poaching of jaguars may be greatest where the population is largest: the Amazon Basin. This reality thwarts scientists as well as environmentalists. In 1980 the field biologist George Schaller, then of the Wildlife Conservation Society, was forced to abandon his jaguar study in Brazil's Pantanal because ranchers there shot both of his radio-collared cats. The perpetrators were never prosecuted, and Schaller concluded that law enforcement was too haphazard for him to continue safely. The genial scientist returned to the United States and assigned jaguar researchers to work in less volatile areas, including Belize. (Before leaving the Pantanal, Schaller broke a toe while chasing a jaguar through the jungle; later he joked that this was his worst injury in more than fifty years of fieldwork.)

"Ranch hands casually kill jaguars, professional hunters guide foreign clients on illegal shoots, and hide collectors take an unknown toll," Schaller wrote of his Brazil experience in *A Naturalist and Other Beasts*. " 'It is impossible to kill off all jaguars,' one rancher told me smugly. 'Some will never be found in the dense bush.' I could have told him that such ignorant words were probably spoken about the passenger pigeon, too, but I merely stressed that the jaguar is already extinct or reduced to occasional stragglers over large parts of the Pantanal, in some areas because of systematic eradication by ranchers. . . . No species in which a female raises an average of only one cub every two years can stand such heavy attrition."

Schaller pointed out that big predators such as jaguars generally exist at low densities and in small populations. Overall, unlike in Africa and Asia, relatively few large mammals inhabit the New World's tropics. Jaguars nevertheless have important ecological functions, influencing the dynamics of ecosystems through their impact on prey, other carnivores, and even vegetation. As a keystone species, *Panthera onca* is presumed to hold many forces in balance. "To help a species endure," Schaller wrote, "we need good science, sound policy, and public support, all of which large carnivores tend to lack."

In the hinterlands of Brazil, ranchers and sport hunters continue their long tradition of shooting jaguars. On private land, expeditions in search of *Panthera onca* are said to be relatively common. Within minutes, a person using the Internet easily can find outfitters willing to accommodate those with a desire to hunt a jaguar. As in other countries, many ranches keep a marksman on staff or retainer for the specific purpose of eliminating jaguars and mountain lions.

The Brazilian-born biologist Sandra Cavalcanti of Utah State University defies the stereotype of the bookish, white-male scientific researcher. Her straight chestnut hair hangs in a pony tail almost to her waist, and her dark eyebrows move in sync with a passionate speaking style. The slender Brazilian once told an interviewer that she became fluent in English specifically because of the groundbreaking work done by her hero, Alan Rabinowitz. She needed to learn the language, Cavalcanti said, in order to read the Brooklynite's memoir about Belize, in which Rabinowitz describes the basics of felid research within the Cockscomb jaguar sanctuary. His expertise helped build the foundation of Cavalcanti fieldwork in the Pantanal, which covers an enormous area. The region encompasses more than seventy-four thousand square miles of flood plain, of which in 2009 about 90 percent was privately owned and host to four million cattle. A hotspot of human-jaguar conflict, the Pantanal may hold the greatest concentration of ungulates in the Western Hemisphere. As a result, some calves and adult cows unavoidably fall prey to jaguars.

But over the years, Cavalcanti's field studies have yielded evidence that few if any Pantanal jaguars feed exclusively on cattle. Her data suggest instead that most actually prefer eating capybara, caiman, peccary, and other game species. Her research echoes Schaller's conclusion that far more Pantanal cattle die from disease, drowning, and starvation than from cat depredation. Even vampire bats, snakes, and vultures may be a threat to such domestic animals as they roam freely through the region.

A study by Utah State's Eric Gese, begun in 2001, documented 436 big-cat kills on private land in the Pantanal, of which 94 were calves and 42 were adult cattle. An estimated 200 of 7,000 head of livestock in the study area were killed by felids, including mountain lions, during the research period. "The livestock kill rate varied from cat to cat," according to Gese. "For some jaguars, cattle were half their diet. For others it was as low as five percent. This was the first time predation rates were documented *anywhere*." Asked what jaguars seem to prefer, in terms of habitat, Gese replied: "To be away from people and where the food is, preferably with some dependably available water."

Such research findings have prompted a few once-skeptical ranchers to pull back on the jaguar hunt. As scientists begin listening to them, the cattle growers start paying more attention to study results. Cavalcanti, who works closely with ranchers and speaks their language, is typical of many wildlife biologists who genuinely sympathize with those who endure livestock losses to predators and want to find alternatives to the eradication of cats. A few biologists also believe that domestic livestock may have physical impacts on the landscape that benefit certain wildlife.

Because of the Pantanal's challenging geography and climate, it has been difficult to arrive at a reliable estimate for the number of jaguars in the region, though as many as 15 percent of all remaining wild jaguars may live there. "No one knows if the population is holding its own or shrinking or growing," the biologist Fernando Azevedo told *National Geographic* magazine's Susan McGrath in an August 2005 article. "All we can say for sure is that ranchers are still killing them." (Azevedo later lost to poachers four of the fourteen jaguars he was studying, including an adult female and her two cubs.)

McGrath noted that "Tonho de Onça" (Jaguar Tony), one of the area's most famous hunters, worked with biologists in tracking and anesthetizing jaguars so that the cats could be radio-collared for research. "That he also still works as a hired gun for ranchers, nobody doubts," wrote McGrath. "I told him I'd heard that ranchers [were] shooting nine or ten jaguars a month in the Pantanal. 'Oh, no,' he said in a vague, airy tone. 'It's much more than that.'"

If this rate of killing is as high as suggested, it raises the question: What happens to so many jaguar pelts? A great number probably are kept locally as souvenirs or trophies, but others obviously filter into the black market. The American University researcher James R. Lee has reported shipments of up to 435 jaguar skins at a time being confiscated from poachers in the Pantanal. Large smuggling rings, concluded Lee, are "set up to export these skins to garment manufacturers in the United States, Germany, and Italy, where finished goods made of the skins are lucrative on the international [black] market."

According to Conservation International, one of several nonprofit groups trying to help manage Pantanal jaguars, the situation has gradually improved since 2000 through establishment of informal reserves and incentive programs whereby ranchers who agree not to kill jaguars are compensated with money, tourism concessions, and medical care. The last is one of the more creative solutions to the poaching problem, wherein rural residents far removed from health facilities are provided access to doctors and clinics as a fringe benefit for helping conservationists. Some cowboys also now supply GPS coordinates for jaguar sightings, kills, and signs, which furnishes researchers with more complete information than camera-traps provide. With the latter, noted Pantanal researcher Eric Gese, "you only find where you look." And at three thousand dollars per satellite transmitter, collaring jaguars is an expensive proposition.

"We have to act," Cavalcanti told the New York Times reporter J. Madeleine Nash in a 2008 report about the urgency of Pantanal jaguar management. Strategies advocated by the Brazilian biologist and some of her fellow researchers include instruction of Pantanal residents in nonlethal forms of predator control and better livestock management, along with payments for jaguar-killed cattle and financial incentives for leaving the cats alone. Some ranches have cut depredation by stringing electric fences, deploying guard dogs, installing bright lights around pastures, instituting regular patrols, and even setting off fireworks at night. At the same time, the New York–based Panthera Foundation and other groups have bought large ranches outright and turned them into wildlife sanctuaries as part of a sideline business in ecotourism. At these locations people are on the hunt for jaguars—but are now armed with expensive cameras and high-power binoculars.

⤙⤚

While publicized hunting safaris fell out of favor following the jaguar's designation as an endangered species, old-timers in Brazil recall fondly the exploits of one daring sportsman who gained worldwide fame in the Mato Grosso wilderness along the border with Paraguay and Bolivia during the 1920s and 1930s. Stories about Alexander "Sasha" Siemel's adventures in the Brazilian jungle thrilled many admirers, who followed his accomplishments avidly through newsreels, articles in such magazines as *Argosy* and *Feathered Shaft*, and a popular 1953 autobiography, *Tigrero!* (Jaguar Man). Born in 1890 in Latvia, Siemel immigrated to South America as a young man and is reputed to have killed more than three hundred jaguars and mountain lions during his celebrated career. Published accounts claim that Siemel fatally impaled as many as thirty-one jaguars with lances or spears, lunging forward to stab the beasts as they advanced toward him.

An avid self-promoter, Siemel played the role of "Tiger" Van Dorn in the fifteen-installment *Jungle Menace* movie series of the 1930s and 1940s, which starred the animal trapper Frank Buck and was recut and released as the 1946 feature film *Jungle Terror*. Siemel also was profiled in a CBS News network TV program and became the subject of a mid-1950s movie project, with John Wayne playing the Latvian hunter. The Wayne picture, costarring Tyrone Power, Ava Gardner, and members of the Karaja tribe, abruptly suspended filming in the Amazon rain forest before its completion. The aborted project became the subject of a 1994 documentary by Jim Jarmusch called *Tigrero: A Film That Was Never Made*. The documentary, which includes intriguing rain forest footage shot as background in 1954, follows the screenwriter and director Sam Fuller as he and Jarmusch return to the jungle village where the screenplay was researched. After laying aside his weapons, Sasha Siemel died peacefully in 1970 at home in Pennsylvania at the age of eighty.

Since Siemel's time, the Mato Grasso has become the nexus of Amazon deforestation, looking in many places more like Iowa than the tropics. Much of its forest has been cut down, often illegally, and turned into grazing pastures and soybean fields. In a telling turn of events, the state governor, Blairo Maggi, headed the largest soy exporting company in the world and sought to challenge the position of the United States as top producer. Maggi's state continues to flood with settlers; their numbers have increased exponentially since the 1990s after the paving of the Cuiabá-Santorém Highway, which now allows quick export of soybeans.

Fragmentation of habitat has made life difficult not only for large cats but for those who study them. "I wouldn't dare give numbers for estimates of jaguars remaining in Brazil," the government field biologist Peter Crawshaw Jr. told me in an e-mail interview. "I would say [the species] is *seriously* threatened outside of the Amazon and the Pantanal, and *threatened* within those two regions. To my knowledge, no killing of a jaguar has ever been legally authorized in Brazil since 1998, but jaguars are frequently killed in retaliation for livestock depredation—usually with no prosecution—because of insufficient and inefficient law enforcement." A 1998 law allows Brazil's jaguars to be killed under the broad category of "defense of property and human life," which is difficult to disprove definitively. I asked Crawshaw what he felt were the biggest threats to jaguars in Brazil. "Direct persecution from ranchers," he replied, "along with prey-base depletion due to direct competition with man as well as habitat modification and fragmentation."

Damage wrought by wealthy political untouchables might be added to Crawshaw's list. In the course of my research, the former Canadian TV journalist Don North recounted to me an experience he had while producing a film about Saudi Arabia on behalf of that kingdom's long-ruling royal family. During the early 1990s North accompanied a Saudi prince, whom he asked me not to name, on a safari in the South American jungles:

> We went hunting jaguars in Paraguay, near the town of Filadelfia. It was probably the most dangerous thing I ever did. The danger was not from the jaguars but from the trigger-happy prince in pursuit of his prey while I was trying to shoot it all on video. We stayed on a big cattle ranch whose owners were known to [the royal family]. Big cats rule the jungle there and

dine regularly on calves. So the ranchers welcome the odd hunt, particularly by the likes of a Saudi prince who pays for the privilege. A couple of good ol' boys were brought down from Texas with their cat-tracking hounds. Over several weeks, the prince probably shot about ten jaguars and pumas. A taxidermist followed the safari, and the prince later built a new wing on his palace to display the mounted cats.

I was unable to confirm North's story with the Saudi royal family, but researchers familiar with the situation in Paraguay told me that it is entirely plausible. Along the Paraguay-Brazil border, it seems, money talks—loudly. Hunting of jaguars is apparently so commonplace along this frontier that one local cowboy admitted to a *National Geographic* writer in 2001 that he had single-handedly killed at least seventy jaguars in the region without ever being prosecuted. However, this job is not without physical risk. In December 2006 a report of a rare jaguar attack on a hunter surfaced in a village fifty miles west of Filadelfia, though it was unclear what the man may have done to provoke the cat. In 2009 sportsmen reportedly were paying thirty thousand dollars each for private (and discreet) jaguar hunts in Brazil and neighboring countries.

CHAPTER NINE

"He Believes He Is a Jaguar"

FOR MANY INDIGENOUS PEOPLES of the Americas, the jaguar is top cat—and always has been. As the unassailable ruler of all animals, in the minds of many, jaguars are closely associated with hunting rituals and the practice of war. In some cultures permission must be received from this powerful feline, through its intermediaries, before game may be killed within the animal's territory. Proof that this cat is a master hunter is believed self-evident in its ability to prowl, climb, swim, stalk, and kill equally well in nearly any environment at any time. The adult jaguar is further admired for meeting its needs without dependence on a social network or family. It gains respect for appearing morally neutral, dispatching victims quickly and without undue cruelty, mercy, or malice.

Such veneration goes back millennia. The burial tombs of many pre-Columbian rulers, shamans, and warriors contain jaguar icons, totems, fetishes, charms, and related materials. Artifacts drawn from gravesites include jaguar teeth and bones, as well as depictions carved in metal, stone, or bone. In some societies the jaguar was viewed as a male symbol imbued not only with great tracking skills and enviable physical prowess but also with superior leadership qualities, persuasive sexuality, and dominance over females.

Tropical America's premier alpha animal, the jaguar has long played a special role in the lives of alpha humans, including religious leaders. These are the witch doctors, herbal healers, and medicine men of tribal cultures and ancient civilizations. Such shamanic figures played a central role in defining the human relationship with jaguars.

Although their repertoire varies, shamans generally believe that they

possess supernatural powers to heal or cause sickness, to summon and communicate with spirits, to alter perceptions of reality, and in some instances
to transform themselves into praiseworthy animals, particularly jaguars.
The latter process is called shape-shifting. A shaman's transformation from
human to animal generally occurs during an altered state of consciousness,
often ecstatic or visionary, that follows specific ceremonies and consumption of naturally occurring hallucinogenic compounds.

Shamans present an enigmatic subject for some Western anthropologists, who might prefer scientific explanations of shape-shifting and other
unusual ritualistic behavior. Researchers sometimes suggest that these charismatic individuals—usually men—are nothing more than nimble-fingered
charlatans or clever magicians. Yet even academic observers concede that
cultural traditions, psychoactive drugs, and abiding faith conspire to play a
central role in a shaman's presumed ability to perform convincing healings
and seemingly supernatural deeds.

The shamanic tradition of the Americas is most visible in tribes of
hunter-gatherers, yet it has occurred within agrarian communities and
complex societies as well. The phenomenon is believed to have originated
among the Mesolithic people of Europe, Africa, and Asia before their prog-
eny came to the New World tens of thousands of years ago. Far from archaic,
shamanism is alive and well. It flourishes among indigenous groups from
Arctic tundra to Patagonian steppes. It survives in the hearts and minds of
millions of Latin Americans of mixed race, demonstrated by the mestizos'
pervasive faith in *curanderos* and *curanderas,* male and female "natural
medicine" healers who rely heavily on native plants, cultural practices,
and wisdom traditions. In some regions, such as coastal Brazil, shamanism
manifests in a synthesis of local indigenous beliefs with those brought from
Africa on slave ships centuries ago.

At the core of the shaman's craft is a vast catalogue of vascular plants,
mushrooms, and a few animals, such as "poisonous" frogs and toads. The
key chemical extracts from these items interact with the human brain and
central nervous system to alter mood, perception, behavior, and experience
of reality. This occurs in sometimes profound and unpredictable ways. Such

compounds are labeled psychotropics, a category of substances that create temporary changes in the brain's "default" cognitive and sensory functions.

About one hundred of these pharmacological tools have been isolated in nature, and more continue to be found and studied by scientists. Most are unknown to the general public and only a few—notably peyote and ayahuasca—have been the object of recreational or sacramental use among urbanites. Peyote is a spineless cactus found from the southwestern United States through central Mexico that prompts psychedelic effects when the mescaline it contains is ingested; ayahuasca is a brewed drink with psychotropic attributes derived from a specific jungle vine and other plants found in the Amazon basin.

In order to increase a psychotropic material's potency, some South American shamans store compounds in hollowed jaguar bones or within medicine bags crafted of jaguar leather. Others, like members of the Amazon's Waiwai tribe, keep hallucinogenic potions close to their bodies as they play sacred rhythms on jaguar-skin drums. They believe that to be like the jaguar—or at least trade on the cat's perceived strengths—is to maximize shamanic power. The jaguar's extraordinary skill and adaptability as a hunter contributes to a widespread belief among tribal people that a shaman who transforms himself into such a cat can accomplish almost anything, whether judged good, evil, or something in between. When a shaman consumes a psychotropic compound, paints his face with spots and rosettes, or adorns himself with the teeth, claws, and skin of a jaguar, that shaman may believe that he or she acquires the ability to rule the forest, expand sensory abilities, and explore all dimensions of consciousness. In some cases the shaman is convinced that he or she literally becomes a jaguar. In other instances the shaman remains human but acquires a set of perceived jaguar abilities and traits. In either situation, the individual gains superior status in the hierarchy of the natural world. Immersed in a heavenly trance and free of the restrictions of society, the shaman believes that he or she can perform extraordinary feats, including healing the lame and afflicted—or exacting revenge on a chosen enemy.

According to the ethnographer Robert Carneiro, use of ayahuasca by Amahuaca shamans of eastern Peru permits them to contact a jaguar spirit that tells the shaman "everything he wants to know, including the immediate whereabouts of the intended victim" of shamanic witchcraft. Among the Yekuana of southern Venezuela, the anthropologist Theodor Koch-Grünberg reported that, while under the influence of ayahuasca, a shaman mimics the roars and other vocalizations of jaguars. Many other examples of imitative behavior among scores of Amazonian tribes are recorded in the field reports of social scientists.

The ayahuasca brew is often used as a tool specifically for jaguar transformation. The first known report of its existence to the modern world was by Manuel Villavicencio in 1858. "This beverage is narcotic," the Ecuadorian geographer wrote, "and in a few moments it begins to produce the most rare phenomena. Its action appears to excite the nervous system; all the senses liven up and all faculties awaken; [those who ingest ayahuasca] feel vertigo and spinning in the head, then a sensation of being lifted into the air and beginning an aerial journey; the possessed begins in the first moments to see the most delicious apparitions, in conformity with his ideas and knowledge."

Social scientists, botanists, and chemists have studied the hallucinogen extensively. It is prepared by boiling together the ayahuasca vine, chacruna shrub, and certain other native plants, then distilling the mixture to a desired potency. The concoction continues to be consumed throughout the Americas for religious and recreational purposes, as well as to rid the body of internal parasites. Like peyote, the bitter-tasting drink is considered a ritual instrument of divination, telepathy, diagnosis, healing, and purification.

Not all purported jaguar-human transformations require the ingestion of the usual hallucinogens. For instance, as reported in the *Los Angeles Times* by Denise Hamilton, some tribal people claim that they get altered eyesight and feel the spirits of jaguars enter their bodies after consuming massive amounts of tobacco. Johannes Wilbert, a University of California,

Los Angeles, anthropology professor, found members of indigenous Amazon cultures who drank up to a quart of tobacco-infused liquid or smoked three-foot-long cigars as a route to trancelike states in which they claimed to foretell the future, travel vast distances as spirits, and transform themselves into jaguars in order to wage war against their enemies.

"Gigantic doses of tobacco produce many changes in the shaman," Wilbert told Hamilton. "They can cause [the shaman's] respiratory system to stop and start, creating the illusion of death and resurrection. They can sharpen night vision or cause dimmed vision and loss of color perception. . . . [Thus a] shaman novice can see at night like a jaguar, he smells like a jaguar, he feels like a jaguar . . . he has no problems believing he *is* a jaguar."

Pelage also is a factor in the ritual role played by *Panthera onca* within indigenous cultures. Among some tribes in South America the black jaguar is considered the most fearsome and powerful predator of all, although knowledgeable scientists insist that melanistic cats are no more aggressive than others. In contrast, the Desana of northwestern Brazil believe that only the "spotted" jaguar cures illness. Among most tribal peoples, coloration appears to carry no particular significance.

Belief in shamanic jaguar transformation is by no means limited to South America. According to Mark Miller Graham and other anthropologists, it survives in Panama and Costa Rica among myths and stories of the Chibcha-speaking people, whose ancient *metates* (grinding stones) were embellished with jaguar imagery and used to prepare certain native plants ingested for mind-altering purposes.

In Guatemala, Maya for centuries have associated jaguars with magic and sorcery. The archaeologist J. Eric S. Thompson describes in *Maya History and Religion* a prevailing belief that a hunter will be more successful if he eats tortillas in which a variety of "jaguar ants" have been cooked, insects so called because their red and black markings are thought to resemble the dark blotches of a jaguar's pelt. "Here clearly," writes Thompson, "eating jaguar ants gives one the prowess of the jaguar, which preys on all game."

In his autobiography, *Secrets of the Talking Jaguar,* the teacher Martín Prechtel describes his initiation during the 1970s as a shaman within the Tzutuhil Maya community of Santiago Atitlán, Guatemala. Prechtel's pedigree made him an unlikely candidate for such a role. He was raised in New Mexico on the Santo Domingo reservation, where his mother—a First Nations (Native) Canadian—was a schoolteacher. The blond-haired, blue-eyed Prechtel—whose father was of European descent—moved to Guatemala in 1971 and was soon chosen by the shaman Nicolas Chiviliu to learn sacred traditions.

The title of Prechtel's book is taken from an experience in which he claims Maya villagers presented him with two jaguar cubs. Prechtel recalls that he returned the young cats to their mother, who then became his oracle of shamanic wisdom. The jaguar's conveyance of divine knowledge was a surreal experience that appears not to have surprised Prechtel in the least.

"To the [Maya] shaman," he writes, "all the places, animals, weather, plants, and things outside in the world are also inside of you as your twin." According to Prechtel, "shamans think that the way one individual goes, so goes a society of individuals. . . . Perhaps the principles involved in some shamanism might be the cause for survival of the collective human spirit in the modern and post-modern era."

There is no doubt that the drug-abetted ceremonies of a shaman alter consciousness. Yet the psychotropic-fueled rituals and transformative experiences of shamans, priest-kings, and other spiritual seekers could not prevent disintegration of the Classic Maya civilization. Among the many theories about the reasons for its collapse, one holds that a series of disasters—droughts, earthquakes, wars, food, and game shortages—shook the abiding faith the Maya held in their religious and political leaders, thus undermining the foundations of society. Even jaguars, important sacrificial animals and portals to sacred worlds, are said to have become scarce.

Although scores of other examples could be cited, one curious practice of the Maya's Late Classic epoch worth noting is the ritual use of hallucinogenic enemas by shamans and royalty, sometimes preceded by the drinking of an alcoholic beverage from a cup shaped like a jaguar's head. The enemas, as reported in the October 1995 issue of *Discover* magazine, probably combined tobacco juice, morning glory seeds, alcohol-laced mead, and psychotropic mushrooms. Delivered through the anus via tubes of hollow bone, these concoctions apparently induced a trance state faster, deeper, and with less nausea than oral ingestion of the drugs. Archaeologist Gyles Iannone, working in Belize, uncovered a ceramic vessel depicting the jaguar deity of Xibalba (the underworld) with a funnel-shaped mouth emitting a "cosmic howl" of ecstasy. Closer examination and subsequent discovery of enema paraphernalia led Iannone to conclude that the set-up was part of a shaman's anus-centered hallucinogen delivery system.

My research on the shaman-jaguar connection led me in many directions, including to a redwood forest close to my then-home. On a Sunday morning, I drove into the Santa Cruz Mountains in order to share brunch with the journalist-researcher Liz Rymland. The author of several scholarly works on the subject, Rymland has interviewed numerous ayahuasca-using shamans from tribal cultures in South America, as well as members of ayahuasca-based religious groups in the United States. The latter are loosely affiliated followers of the Santo Daime religion, whose right to ingest the hallucinogenic brew as a holy sacrament has been upheld by the U.S. Supreme Court. Of particular interest to me were the transcripts of Rymland's extended conversations with shamans and tribal elders in the rain forests of Ecuador. She had spent time among them as an activist working for nonprofit groups trying to preserve shamanic traditions. Over an omelet and hot tea, Rymland shared her belief that the breakdown of hunter-gatherer societies in Latin America—through habitat destruction, Christianization, disease, alcoholism, and cultural encroachment by Westerners—was fast eroding an irreplaceable set of human-jaguar connections.

"Within many traditional indigenous communities, especially in Ecuador among the ethnic minorities, the elders who embody the ancient ways and knowledge are fewer and fewer," Rymland told me, as morning fog burned away. Among a dozen ethnic minorities in the Ecuadorian Amazon, she said, there were probably fewer than ten knowledgeable old-timers still alive within each group. In Brazil, entire rain forest tribes continue to be wiped out or assimilated, often without documentation of their beliefs and traditions.

Rymland was concerned that a fragile connection to age-old wisdom was being broken irreparably. "Unless this bridge can be widened," she said, "priceless and irreplaceable knowledge of the uses of plants and of our interrelationship with nature will not survive." While some tribal shamans still knew venerable jaguar-transformation rituals, Rymland cautioned that many of these men were old and frail. They had not passed their specialized lore on to younger community members, nor had outsiders preserved such knowledge for posterity.

"They truly believe they are transformed into jaguars," Rymland said of shape-shifting shamans, responding to my final question. "And if you see them in that state of mind and body, as I did, you will swear it is true."

Some anthropologists believe that the remote lowland tropical basins of South America are where the jaguar is most closely associated with sacred tribal ceremonies and origin stories. According to members of the Amazon's Tukano tribe, for example, the sun god created jaguar as his representative on Earth. The cat was bestowed with the yellow color of sunshine and the growl of thunder, which was said to be the sun's voice. A Tukano legend holds that Earth's very first man found a sacred trumpet, from which music flowed to make the stars, wind, rivers, fish, forest, game, and his fellow humans.

Members of the Matsés tribe, who dwell in the Río Gálvez rain forest of Peru and Brazil, surprised European "discoverers" in 1976 with vivid facial adornments intended to resemble and pay homage to the jaguar. These include thin bones piercing the flares of their noses that equate to cat whiskers, shell earrings said to look like a jaguar's ears, sticks puncturing lips to evoke fearsome canine teeth, and tattoos or dye imitating the felid's

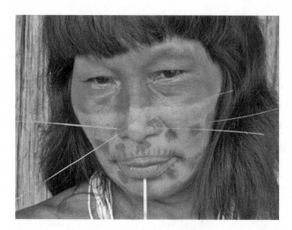

Figure 9.1. A female member of the Matsés tribe of the Amazon basin adorned in a traditional fashion intended to emulate jaguars, which the Matsés people revere highly. Nose reeds replicate a jaguar's whiskers, while face paint echoes fur patterns. Photo © Dan James Pantone, www.Amazon-Indians.com

spots, stripes, and mouth. Sometimes called "the cat people," the Matsés became well known in the late twentieth century after the British explorer Benedict Allen learned survival skills from them during an epic thirty-six hundred–mile journey. He describes the tribe vividly in his autobiographical 1994 book, *Through Jaguar Eyes: Crossing the Amazon Basin,* relating how Matsés are divided into symbolic clans named after various jungle animals, including the jaguar. Masters of 12-foot-long blowguns, the tribe's most adept hunters reportedly can strike a hummingbird in flight with their poison darts and a monkey at 130 feet. The paste for these darts is occasionally mixed with jaguar hairs, said to give the venom extra potency.

As described in Mark Miller Graham's *Reinterpreting Prehistory of Central America,* there are innumerable instances of other tribal groups enshrining the spotted cat in cultural practices and spiritual beliefs. The wearing of jaguar teeth and skin has been particularly common, along with dancing that imitates the cat and employs masks mimicking its face. A random sampling of such tribal customs includes:

- Brazil's Suya prescribed jaguar meat for every category of age and gender in the tribe, but also hunted the cat for its claws, teeth, and hide, which were used variously in necklaces, dancing belts, and trade items.
- The war-loving Bororo of southern Brazil believed that a jaguar should be killed by the spirit of a dead person when the time was right to return human society to its correct balance.
- Hunters of Panama's Kuna tribe killed a jaguar in order to burn its heart and eat the ashes, which they believed made them skillful hunters. The accomplished Cubeo jaguar hunters of the Amazon, in turn, wore jaguar-teeth girdles as symbols of their own expertise.
- Members of the now extinct Héta tribe of Brazil wore jaguar skins and hats at special communal ceremonies, while the Waiwai of Guyana played sacred jaguar-skin drums on celebratory occasions.

Although I had direct experience neither with ayahuasca nor with the tribal people of Amazonia, I encountered several Maya shamans over the years while exploring Mesoamerica. I repeatedly noticed sacred jaguar-shaman imagery among ancient ruins and museum collections in Mexico, Guatemala, Honduras, and Belize. But my most powerful set of "sacred jaguar" experiences occurred in the fast-disappearing subtropical forest that lies in the borderlands between Guatemala's Petén and the Mexican state of Chiapas.

In March 1994 I joined a whitewater raft expedition down the remote Río Usumacinta to investigate some of the Classic Maya's most impressive spiritual centers. In conjunction with this trip, I also visited strongholds of the Lacandón tribe, some of whose members are the most traditional of the region's Maya.

I accompanied a two-person crew led by outfitter Scott Davis—a cheerful, ruddy-faced outdoorsman who could charm the eyelashes off a rattlesnake—at a put-in point in Corozal Frontera, Mexico, to begin our five-day float down the Usumacinta. The waterway's name means "the sacred river of monkeys," and it is wide and impressive, draining a lush watershed of hundreds of square miles as it flows northeast toward the Gulf of Mexico. But we were to see only a few of the black howler and spider monkeys that once were numerous along the Usu's banks. Davis explained that all wildlife had diminished over the years, presumably owing to encroaching settlement in Mexico and Guatemala, a portion of whose international border is formed by the river. Along our route there were still no roads along or across the Usumacinta, but they seemed inevitable.

In late afternoon of our first day, Davis was pleased to point out a flock of scarlet macaws, soaring en masse from one country to another. This large multicolored parrot is highly valued in the black-market wild pet trade. Throughout Mexico, where it once ranged nearly to the Texas line, only a few hundred remained. The three-foot-long birds made loud, low, throaty squawks, screams, and squeaks as they cavorted through the air. It was a wondrous spectacle.

"Maybe ten years ago we'd have had a decent chance of seeing a jaguar along the riverbank—or even swimming across the river," Davis told me, as he drew back on oars that steered us downstream. "In fact, during the days of the Classic Maya, jaguar skins were one of the top trade items ferried along this river in cargo vessels."

The Maya traders had to be accomplished sailors. We saw few other boats on the Usumacinta, which is still avoided by many because of its strong currents and challenging rapids. Our nights were spent on beaches wedged between the dark forest and moonlit water. It felt as though humans had never lived here—and that the Maya had not existed. Yet we knew that there were smugglers, bandits, and political malcontents hidden in the forest. In the golden hours of dawn, the jungle revealed itself in full glory: with croaking toads the size of hubcaps, crimson orchids dangling from tiny pockets of soil tucked into tree branches, and gymnastic monkeys chattering their greetings.

Our own trip down the Usu proceeded without incident, though we had M-16 semiautomatic rifles aimed at us briefly when we landed at Piedras Negras, a Maya archaeological site then occupied by Guatemalan guerrillas. For forty years, members of this insurgency had been locked in armed combat with their militaristic government. But the rebels had no intention of harming us. Instead, they requested that we hear a twenty-minute lecture outlining their political grievances, after which they would allow us to explore the overgrown ruins on our own. The lecturer faced us stiffly as he delivered his speech, but upon completion set his rifle aside and lit up a hand-rolled cigarette. He even offered to guide us into the heart of the ruins, which had been ignored by archaeologists and tourists for many years.

We were awakened the next morning by gunfire from target practice at the nearby guerrilla encampment. Needless to say, no jaguars or other shy animals revealed themselves. Although this unstable political situation made the river corridor a problematic place for humans, it perhaps improved survival odds for the cats that remained hidden. Here as elsewhere, a rain forest patrolled by armed rebels and criminals curtailed settlement—and, theoretically, habitat destruction. Cutting of the forest rebounded, however, following the signing of peace accords in Guatemala in late 1997 and a concurrent calming of Zapatista guerrilla activity on the opposite bank in Mexico. (The leftist rebels—named after the celebrated Mexican revolutionary, Emiliano Zapata—were engaged in a long-running political struggle against Mexico's federal government and had seized briefly the highland city of San Cristóbal de las Casas.)

Over dinner on our second night at Piedras Negras, campfire conversation turned to speculation about what had caused the spectacular collapse of the Maya civilization. Our group included the Santa Fe writer-photographer Don Usner, a thoughtful naturalist who had traveled extensively in the Maya world. He and other members of the party focused on a variety of plausible factors, including destruction of a fragile ecosystem that on its surface appears fecund and lush. But tropical forests generally have shallow, swiftly depleted soils, and farming them is not easy. The pyramid of animals competing for survival is kept strong through a complex web of interdependent forces, including the presence of keystone species like the jaguar. Tampering with that web can unleash a cascade of catastrophic events that eventually undermine cultural traditions and the natural cycles of the forest.

I noted research suggesting that so many jaguars were killed for sacrifice at the large Maya city of Copán, in present-day Honduras, that its rulers were forced to substitute less desirable mountain lions and ocelots in bloodletting rituals. Archaeologists theorize that Copán's priest-kings became increasingly eager to please their deities with the blood of sacred animals, particularly the jaguar, but at some point they simply could find no more. During a later visit to the site, I was told that the cats had returned, glimpsed occasionally by visitors touring the ruins.

Certainly, a scarcity of jaguars and other forest dwellers did not make the Maya superstructure wobbly enough to crash. But it may very well have been symptomatic of a civilization in which an excess of royal families— jockeying among themselves for status and position—depleted the available supply of an animal deemed essential for ceremonial bleeding, status-conferring clothing, and mortal sacrifice.

"Any society that depends on growth economics, with elites reaching ever-greater levels of material well-being, eventually reaches its limits," observed the Vanderbilt University archaeologist Arthur Demarest in a 1990 *Los Angeles Times* interview. "The balance between ecology and society is exquisitely delicate. If something throws that delicate balance off, it all can end." Demarest, a specialist in Late Classic Maya cities of Guatemala's Petén forest, including Piedras Negras, cited evidence to support his theory that ecological disaster befell the Usumacinta basin around A.D. 900.

Our group toured the monumental Mexican ruin of Yaxchilán, which overlooks an oxbow in the river. The imposing Classic Maya site was believed dominated by a "jaguar cult" dynasty, beginning with a ruler named Yat Balam (Penis Jaguar) in A.D. 320 and extending in the seventh and eighth centuries through the reigns of Escudo Balam (Shield Jaguar) and his son, Pajaro Balam (Bird Jaguar). Clinging to the hills were magnificent carved stone reliefs and hieroglyphics depicting Yaxchilán rulers wearing adornments made of jaguar skins and other body parts. One carving showed an adult female presenting a jaguar mask to a priest; another depicted a woman drawing sacrificial blood by piercing her tongue with a barbed string. Our group arrived finally at Building 33, perhaps the most famous structure in Yaxchilán, where we stood before a decapitated limestone sculpture of Bird Jaguar. The Lacandón Maya say that when the head, which lies only a few feet away, is returned to its place atop this royal figure, the world as we know it will be destroyed by "celestial jaguars." Some predict that this prophecy, which is said to portend a new creation cycle, will be fulfilled on December 21, 2012, a date on the Maya calendar when a long epoch has been forecast to end.

During our exploration of the Usu watershed, the river guide Scott Davis took us to Lacanjá, a village of traditional Lacandón who live separately from a Christianized faction of the same tribe. Here I interviewed a young man named Enrico Chan K'in, who seemed content to straddle worlds old and new. He wore short hair, a T-shirt, jeans, and flip-flop sandals rather than the white, ankle-length, single-piece gown and long hair of traditional Lacandón Maya. Only a few minutes before our conversation began, I had stumbled upon a scene that disturbed me: a hunter was sharpening a knife in preparation for killing and butchering a terrified spider monkey, which struggled in vain to escape a tight set of restraining ropes.

Still dazed from what I had just witnessed, I asked Chan K'in, "What role do the larger animals of the jungle, such as monkeys and cats, play in your lives?"

"We need to eat them in order to survive," he replied, "but we also worship these creatures. You see, every living thing is connected, from the smallest ant to the biggest jaguar. We humans are part of that connection as well." Chan K'in explained that monkeys are an inferior version of men, in the Maya worldview, because they are the result of a failed attempt by the "creator god" to make people. "But the jaguar," Chan K'in added, without prompting, "is not a failure. It is a complete success. We respect him above all other animals in the forest."

I asked about the elders and shamans of the traditional Lacandón and was told that their power had diminished substantially. While still respected, their authority was marginalized by outside influences.

Later that day, Davis led us to Bonampak, a ruin in a small clearing about seven miles by trail from Lacanjá. A Classic Maya colony controlled by Yaxchilán's ruling jaguar dynasty, this modest city was "discovered" by a Western photographer in 1946. The Lacandón had been praying among its temples for centuries. They still do.

Of greatest interest at Bonampak are its world-famous murals, painted on stucco surfaces inside three windowless buildings near the top of a sixty-foot limestone block pyramid. Although these precious paintings are now sealed off for their protection, at the time of our visit they could be viewed close up, even touched. Unlike most Maya frescos, those at Bonampak were shielded by clever design from the elements and further protected by the accidental leakage of calcium carbonate, which formed a shield from decay. The murals document, with great detail and realism, a series of events related to the last ruler of the city, which was abandoned around A.D. 800.

The buildings sheltering the artworks are surprisingly small, and we had to take turns squeezing inside. I contorted myself like a pretzel in order to see sections of the paintings spread across each ceiling and wall. The first room showed the clothing and adornments of various priests and nobles gathered for a high-court ceremony marking Chaan Muán's ascension to the Bonampak throne. Also visible was an orchestra playing wooden trumpets, drums, and other instruments, while members of the nobility danced and conferred with one another. The nobles' costumes included a vivid combination of jaguar pelts, quetzal feathers, and boa constrictor leather.

The second room displayed an elaborate battle scene in which, after violent hand-to-hand combat, prisoners are taken, with bleeding fingers, to be seated before the richly attired Chaan Muán. The ruler is seen grasping a captive by the hair while holding in his opposite hand a spear decorated by a jaguar pelt. An elite warrior wearing a jaguar tunic and fancy headdress accompanies Chaan Muán. The last room re-creates a ceremony with dancers wearing jaguar-themed costumes and masks honoring various Maya gods, while the new ruler and members of his family stick stingray needles and barbed strings through their tongues for ritual bloodletting.

The thrill of admiring these masterpieces—perhaps the most vivid among only a handful of such murals remaining—enlivened our conversation during the two-hour slog through dense, mosquito-infested jungle back to Lacanjá.

"What did you think of the frescoes?" Enrico Chan K'in asked me upon my return to his village.

"They were very special," I told him. "I particularly enjoyed seeing so much clothing and ornamentation made from jaguar pelts and body parts."

Chan K'in grimaced and exhaled a sigh: "We almost never see *el tigre* anymore. With so many people in the forest, they have gone away. I wish we knew where. I'm afraid that we have lost the protection with which they blessed us for so many generations."

A couple of weeks later I debriefed my Guatemalan friend Victor Perera about the trip and outlined my conversation with Chan K'in in Lacanjá. Perera, who died in 2003, was a gifted poet and the coauthor of *The Last Lords of Palenque,* a lyrical book about the Lacandón Maya culture. Perera floated the Usumacinta with Davis several years before I did and was discouraged by my updated report of habitat destruction. He estimated that, based on scientists' estimates, the Lacandón rain forest had shrunk from 90 percent of its original size in 1965 to less than 30 percent three decades later. "This unregulated encroachment, fueled by population growth, may yet sound the death knell for what remains of the Lacandón jungle," said Perera. "This could eventually cause an ecological chain reaction of devastating proportions."

After bidding adios to Scott Davis and my raft-trip friends, I boarded a crowded public bus for a long ride into Mexico's Sierra Madre Sur. I was writing radio, magazine, and newspaper stories about the threatened Usumacinta ecosystem and wanted to conduct a last round of interviews in

San Cristóbal de las Casas. We were stopped at one roadblock after another, each swarming with nervous teenage soldiers. The Zapatista rebellion had brought the Mexican army out in full force, and I was searched—along with other male passengers—several times before our crowded bus reached its destination.

Tourism had been hit hard, and swarms of children surrounded me in the otherwise empty Plaza Central. Most were selling some variant of a thumb-sized Zapatista guerrilla doll that wore a tiny black ski mask and held a miniature rifle, emulating the rebel leader Subcomandante Marcos. I bought a few as souvenirs, then made my way to Na Bolom (Jaguar House), a nonprofit independent research center long associated with the indigenous peoples and natural resources of the Chiapas rain forest. I interviewed Na Bolom's director of cultural studies, Will Hoffman, who allowed that "environmental education [was] desperately needed" among the thousands of homesteaders streaming into the lowland jungles.

"There is much that these newcomers can learn from the more traditional Maya," Hoffman said. "The Lacandón have farmed successfully for hundreds of years with minimal disruption of forest ecosystems." Such disruption, I assumed, was behind Enrico Chan K'in's observation that jaguars had grown scarce in the Lacandón rain forest.

After a few days I continued north to the Chiapas capital of Tuxtla Gutiérrez. With several hours to kill before the departure of my flight to Mexico City, I took a taxi to the Manuel Álvarez del Toro Zoo, named after its naturalist founder and considered one of the best such facilities in Mexico. I joined the throng admiring a male black jaguar ensconced behind glass in a spacious enclosure. The captive cat was named Sombra, Spanish for shadow. He paced restlessly along the cage's perimeter. Although the animal appeared physically healthy, his confinement in such a small space left me feeling glum and frustrated. I sat on a bench opposite Sombra and contemplated the intertwined fates of jaguars and people.

As I stared at this captive cat, it occurred to me that something vital had been lost in our society's transition from its hunter-gatherer origins to today's highly individualized and technological civilization. Perhaps I had drifted into naïve romanticizing, but I felt that the intimacy with nature so essential to tribalism had morphed into a dissociation bordering on indifference. Somehow, humans had traded their precise knowledge of the natural world for an almost completely mediated environment. Jaguars were better known as cartoon characters or hood ornaments than as flesh-and-blood animals. The shaman, who gains insight by using organic compounds to explore consciousness, was slowly disappearing from the New World. It seems that we have few other sources to remind us of what is sacred within the plants, animals, and phenomena of nature—except the natural world itself. But on a fast-urbanizing planet, that experience is becoming more abstraction than reality. (In 2006 I was saddened to learn that Sombra had died of a tranquilizer overdose in what should have been a routine caretaking procedure.)

CHAPTER TEN

"There It Is; I'm Going to Shoot It"

I WAS OFF AGAIN TO BELIZE, the little country believed to have one of the largest densities of jaguars in North and Central America. I still sought a wild jaguar, but I also wanted to investigate disturbing reports that big cats in this Massachusetts-sized nation—which I had visited fourteen times in as many years—were being killed at an alarming rate. Emerging here was the same set of interrelated problems associated with other, more crowded, developing regions.

I first spent a few days in the forested foothills of the Maya Mountains, where several ecolodges promised exciting outdoor escapades for their guests. Day-trip choices listed in one brochure included rafting, rope-climbing, and "tubing" in nearby caves, as well as jungle treks in search of parrots, monkeys, and other exotic wildlife. I made a halfhearted attempt on my own to find a jaguar in the Caves Branch area, a patchwork of forests and orchards where I was told the cats are sometimes seen, but found no direct evidence of their presence. The local residents I interviewed, all recent arrivals from Guatemala and El Salvador, said most wild game had disappeared as the human population had increased. Under pressure from hunters and without sufficient prey, the jaguars had apparently moved on —or starved.

"On a good day," offered one Guatemalan immigrant picking fruit in an orange grove, "we can earn twenty-five dollars. On weekends, we go hunting."

A high point of my visit was a day spent with a mestizo guide named Vida inside Footprint Cave, one of many chambers in the area formed by underground streams. Following a jungle-shrouded waterway that disappeared into the side of a hill, we approached the cave by alternately walking and paddling our way upstream. Deep inside this ancient Maya ceremonial site Vida showed me a fist-sized carving he said represented the Classic-era rain god Chac, one of the most important deities in the Maya pantheon. Below this effigy, cut into a limestone bench and emblazoned faintly with peeling pigments, were representations of an erect penis and an extended tongue, marking the place as a sacrificial altar for ritual bloodletting.

Leading me a half-mile farther into the cave, my guide showed me ancient pottery, fire pits, and a half-dozen footprints preserved in rock-hard calcite. The prints appeared as though they had been left only days earlier, yet a thousand years was a better estimate. Nearby stood a large pot from which archaeologists had scraped centuries-old corn liquor. The vessel bore a painting of a human figure in a bowed posture, probably prostrating himself before an underworld god or deceased priest-king.

"Listen," Vida commanded, placing an index finger against his lips as we sloshed our way back to the mouth of the cave. "Do you hear the whispering?" When I closed my eyes, the soft sounds of running water echoed off low, smoothly eroded walls to replicate hushed voices in secretive conversation.

"The ordinary Maya were afraid to come in here because they believed those sounds were spirits of the dead speaking from Xibalba," said Vida. "The royalty who entered wore face-paint in order to scare the ghosts away." I shot my guide a skeptical look, but he assured me that he was merely passing along information that visiting archaeologists had told him.

When I asked Vida about flesh-and-blood jaguars, he shook his head and said that such cats were gone. "During my lifetime," sighed the twenty-four-year-old, as we floated on our inflated car-tire inner tubes toward the shadowy entrance, "they have all disappeared."

I was confident that jaguars had made regular use of such grottos as Footprint Cave before they were rediscovered during the 1930s by itinerant gatherers of chicle, a tree sap that was once an essential ingredient of chewing gum. But since the 1990s this place had joined Belize's established tourist circuit. Jaguars would no longer feel safe here.

The next morning I hiked to Five Blues Lake, a picturesque forest-enclosed lagoon where I encountered several peccaries, the favorite local prey of jaguars, but no cats. The lovely lake had plenty of birds, as isolated scraps of jungle often do, but seemed largely devoid of ground-dwelling animals. As elsewhere in Belize, what was once a remote parkland had been made accessible by new roads and was now dotted with agricultural plots, many of them producing fruit, beans, squash, and corn. There was plenty of livestock, too, which no doubt set the stage for potential conflict between humans and jaguars.

The following afternoon I settled in at the Tropical Education Center, a research station east of the capital city of Belmopan in rolling, tree-studded savanna. This grassy landscape is indented with waterholes that are home to crocodiles, turtles, frogs, and waterfowl. Clumps of forest provide the kind of cover preferred by ambush-hunters like jaguars. My guest quarters consisted of a commodious *casita* perched on stilts above a gnarly thicket of palms and pines. Screened windows allowed breezes to flow through this aerie, from which I could scan much of the eighty-two-acre reserve. The cabin's interior felt made-to-order: a jaguar-themed bedspread, jaguar-print pillowcases, a jaguar-adorned rug, framed photos of real jaguars, and books about jaguars lining the bookcase. I felt as though I had died and gone to heaven.

On the second morning of my stay, TEC manager Celso Poót appeared at breakfast eager to answer my questions. A full-blooded Maya with a round, kind face and soft hands, Poót told me that three jaguars had been seen on the grounds during the previous year. He theorized that the cats were transients that used the intentionally well-protected property to follow migratory prey, such as peccaries. The refuge abutted the Western Highway, which bisects Belize and upon which vehicles traveling at high speed regularly strike wild animals, including felids. The seasonal migrations persisted, though this busy thoroughfare was a serious barrier to wildlife movement.

"There are many people relocating into our area," Poót said, "and ranchers are complaining more often about 'problem jaguars.'" This term, he explained, was a euphemism in Belize for jaguars suspected of devouring calves, dogs, pigs, and other domestic animals. A government survey counted ninety reported jaguar attacks on livestock during 2003 and 2004, with a suspected increase in subsequent years. In response, between 2002 and 2004 humans had killed an estimated sixty jaguars in Belize. The evening before my arrival a local resident had relayed word to Poót of yet another suspected cattle killer, this one along the nearby Sibun River. If the jaguar was not captured soon, the rancher warned, he would shoot it dead.

Poót ended our discussion on a hopeful note. He recalled that when a livestock-killing jaguar had been captured and removed in southern Belize, local residents expressed concern about the cat's welfare. "'What will happen to it?' they asked. To me, this shows our citizens still care about jaguars."

TEC's co-owned Belize Zoo was working with the national government and Belizean ranchers in a nonviolent conflict resolution program whereby problem jaguars were trapped and transferred to a holding facility across the highway. Instead of being summarily killed, such animals would have their lives spared, habits studied, genetic material dispersed, and role expanded as environmental educators in overseas zoos and wildlife parks. They would not be returned to the wild, on the assumption that jaguars accustomed to killing domestic animals would not readily mend their ways.

"This is a difficult conservation issue," explained Poót in a firm voice. "But it must be addressed in an urgent manner. The idea is to habituate problem jaguars to a new life in captivity. We do this by keeping them off-exhibit in the back of our zoo." The program appeared to be working. Graduates of the "rehab" unit were placed successfully in overseas zoos beginning in 2007. An official of the Forestry Department, which oversees Belize's parks and wildlife, said that the new approach helped mitigate the difficult matter of "problem jaguars" but that complaints from disgruntled rural residents were nevertheless escalating.

Figure 10.1. Students from Belize's St. John's College, including several youths descended from the region's original Maya, are taught by the wildlife biologist Carolyn Miller to identify tracks and other roadside jaguar signs. Photo by Carolyn M. Miller

From the Tropical Education Center it was a four-hour drive to La Milpa Field Station in the 250,000-acre Rio Bravo Conservation and Management Area. The wretched, unpaved road angled west from Orange Walk through sugarcane fields and citrus orchards. In avoiding potholes big enough to swallow Volkswagens, I rarely took my rented pickup above twenty miles per hour. At the Rio Bravo, which drains an upland plateau, I peered past flooded rice paddies toward Mexico and, invisible in the haze, Guatemala. The roadway improved when asphalt appeared as if by magic. For this I could thank local Mennonites.

The northwest corner of Belize is home to thousands of prosperous homesteaders united by strong religious beliefs and an enviable work ethic. Like the Amish with whom they are loosely affiliated, Mennonites are a small Christian denomination whose most tradition-bound members have pacifist leanings, shun modern conveniences, and dress in a modest, old-fashioned manner. They ride in horse-drawn buggies, plow with mules, and light their homes with oil lamps. (In Belize, Mennonites include some "progressive" sects who are indistinguishable in appearance from others in mainstream society.) All are part of a diaspora that began in Europe more than four centuries ago and has brought thousands of German-speaking adherents to Latin America's backcountry.

Beginning in the 1950s Mennonites chopped down much of the northern Belize jungle and replaced it with tidy farms, orderly ranches, and bustling furniture factories. Fields of corn and pastures of cattle now exist where two hundred–foot trees once grew. It is a remarkable accomplishment, yet the resulting changes in flora and fauna are dramatic—and perhaps irreversible. They illustrate the larger problem of shrinking habitat that affects not only jaguars but all wild animals requiring large tracts of undisturbed native vegetation. As world population expands, the pressure to fragment these lands for human use becomes enormous.

The naturalist George Schaller once spoke to this international conundrum before an audience in India dedicated to saving wild tigers. "Conservation must be integrated with the needs of people and cultures," he conceded. "[But] society must respect humanity's need for wild places, for the pristine, for places where pillaging is forbidden. . . . Science today must defend: must act as well as gain knowledge."

My destination was a research station inside the Programme for Belize nature reserve, which preserves a huge chunk of La Selva Maya (the Maya jungle). I crossed into the property after passing a patch of adjacent tropical forest in the process of being cleared. A single white-tailed deer stood motionless nearby, wide-eyed and presumably confused by chaotic heaps of slash lining the road. I stopped and took a photo as it bounded away. The deer ran in a zigzag pattern, searching in vain for the safety of a thicket or copse of trees. On the opposite side of the road, ranchers had cleared every shred of vegetation up to the barbed-wire fence marking the PFB property line. Only ankle-high grass and knee-high stumps remained. The contrast between developed and protected land could not have been more stark. Over dinner that evening, my conservation-minded guide at La Milpa Field Station shook his head upon hearing my story: "If that deer stays out there, it'll be shot. Bet on it."

La Milpa is a compound of rustic clapboard bungalows encircled by nearly pristine jungle. It straddles an intersection of unpaved roads that continue south to Gallon Jug (another privately protected area, named after a Maya artifact found there) and west to the ancient ruin called La Milpa (translated as "the jungle homestead"). The Rio Bravo Conservation Area, as this tract is formally known, comprises fully 4 percent of the country's total land area. Its assemblage of biodiversity includes nearly four hundred species of birds, two hundred species of trees, and seventy species of mammals. All told, a dozen endangered species flourish here. Besides research and ecotourism, the nonprofit organization managing the forest carries out sustainable harvesting of timber, chicle, thatch, and palm leaves. A perennial question is whether these efforts can deliver enough revenue to keep the conservation area intact over the long term.

I arrived on a hot, sultry April afternoon. Several young women—ecology students from England—sunned themselves on the basketball court, working on their tans a day before returning to a chilly British spring. I settled into simple guest quarters furnished with ceiling fan, mosquito net, and lumpy mattress. I was advised that diesel-generated electricity was available only a few hours each day and that basic meals were served in a communal dining hall.

My escort was Vladimir Rodríguez, a native Belizean with thick black hair and a milk-chocolate complexion. He greeted me with binoculars in hand and pointed in the direction of a keel-billed toucan, the so-called Froot Loops bird, known for its incongruously large, rainbow-colored bill and jaunty personality. The comical bird was carefully picking round nuts off a cohune palm and cracking them with its boat-shaped beak. Vlad's sharp, dark eyes constantly surveyed his surroundings in search of fauna. He had worked at the station for six years and had seen fourteen jaguars. "I've come across them at all times of the day and night," Rodríguez deadpanned, as if describing encounters with a squirrel. "Most were walking on the main road you drove in on, in early morning or at twilight." The champion birder seemed more proud of his success in checking off most of Belize's five hundred–plus avian species from his "life list." Many of these birds had disappeared from the rest of Central America but remained abundant in La Selva Maya, of which Rio Bravo was a part.

Rodríguez introduced me to Ryan Phillips, an earnest, fair-haired young California scientist tracking several radio-collared harpy eagles reintroduced to the area by the nonprofit Peregrine Fund in conjunction with the Belize Zoo. (In 2008 six were observed in the region.) Like *Panthera onca,* the harpy is an apex predator with low densities throughout its shrinking range. Bringing it back to the woods would help balance regional ecosystems. This magnificent creature—its gray-and-white body topped by an ash-gray head with an erectile crest of long feathers—is the second-largest eagle in the world. An average female weighs seventeen pounds and has a wingspan of more than six feet. The birds compete with jaguars for coatis, spider monkeys, and other small mammals. Harpy eagles have almost disappeared from the region, partly because of habitat destruction and misconceptions among the general public. Some rural residents regard them as dangerous or "evil omens" and kill them on sight, saving the eagle's feathers as a talisman. Thanks to education efforts by the Belize Zoo and others, the harpy is shedding this problematic image and becoming an important conservation symbol.

While remaining hopeful about improvements, Phillips was generally cynical about the state of conservation in Belize, noting that the harpy eagle program was moved out of Chiquibul National Park because poachers already had killed so much wildlife. "The eagles simply couldn't find enough to eat because humans took their food," the scientist told me in a dejected tone. His cheeks flushing red, Phillips maintained that only a decade earlier, Chiquibul had been one of the least disturbed forests in the country, occupying prime jungle habitat in the Maya Mountains.

The raptor expert blamed much of the devastation on heavily armed *xateros,* men who roam the bush looking for xate palms, a native understory plant whose delicate fronds are exported by the tens of thousands for decorative use worldwide as "green filler" by commercial florists. Most freelance harvesters were impoverished Guatemalans, entering Belize illegally across a largely undefended frontier. "Xateros come [to Belize forests] for six months at a time and kill everything that moves," declared Phillips.

The scientist's claim was substantiated later during my conversations with government officials and other scientists, a few of whom believed some xateros were involved in drug smuggling and armed robbery, which also plague the region. Nine months later a member of the Chiquibul National Park Advisory Council contacted me and said the xatero situation had grown much worse. This source confirmed that illegal hunting had caused "a dramatic decrease in wildlife of all types" and asked rhetorically: "What are the jaguars doing as food declines? We do not know." (Certification programs by Rainforest Alliance and other organizations seek to reduce the negative impacts of such xate harvesting on the environment, but results of their efforts have been mixed.)

When not interviewing field station naturalists, I spent as many waking hours as possible in search of a jaguar. Phillips had seen fourteen during his several months at La Milpa and showed me magnificent photos. In one picture, two jaguars lolled together lazily in the middle of a road, looking half-asleep. In another, a jaguar faced the camera wearing an obstinate expression, as if challenging the photographer to step forward.

I followed advice proffered by Phillips and Rodríguez, ambling quietly along the park's main roads and trails each day, at dawn and dusk as well as many hours in between. I paid particular attention to the little-traveled road to the ruin, which snaked through several miles of high-canopy bush. I spent hours walking to and from the archaeological site, sometimes bush-

Figure 10.2. A healthy jaguar on the prowl in a Central American jungle, display-ing the characteristic coloring, rosette pattern, and muscular physique of *Pan-thera onca*. This is the same cat shown dead in Figure 10.4. Photo by Carolyn M. Miller

whacking through the jungle on side trails. I was rewarded with spectacular scenery and an occasional wildlife sighting: howler and spider monkeys, kinkajous, coatis, agoutis, armadillos, and deer. There were plenty of colorful birds, including parrots, parakeets, toucans, trogons, tinamous, motmots, and oropendulas. According to another visitor, I just missed a mountain lion at the ruin by a matter of days. But I found no evidence of jaguars. Phillips was empathetic—"More of them are around during the rainy season," he said—while Rodríguez waxed philosophic: "It's impos-sible to predict where or when jaguars are going to appear."

One day, determined to give it my best shot, I searched for thirteen straight hours in the surrounding forest. At one point I stumbled into a fringe area of the La Milpa ruin and startled a troop of coatis, their two-foot tails held high as they rummaged beneath a fruit tree for ripe snacks. These golden-brown mammals are members of the raccoon family and, like these cousins, have black masks and indistinct rings around their tails. Coati bands are generally composed of females and their offspring, trolling in groups of thirty or more with their sharp claws and long snouts. They grub for insects, reptiles, tubers, rodents, and whatever else the jungle might offer. Highly social, they are fascinating to watch, and I spent my lunch hour doing so before moving on to inspect some of the deep pits and long tunnels that looters had left after pillaging La Milpa decades earlier.

During my circuitous journey back to the compound I came upon a fresh set of jaguar tracks along a muddy section of game trail. The impressions were deep and well etched, probably no more than a half-day old. The cat might have left them during an early morning hunting expedition. I followed the footprints until they disappeared into the jungle, then waited for nearly two hours in the vain hope that the jaguar would reappear. Mosquitoes, flies, and gnats got the best of me, and I headed for my bungalow.

"*No suerte, amigo?*" Rodríguez inquired, receiving me at the mess hall that night. He read the disappointment on my face. Phillips slapped me on the back and reminded me of an important reality: "The animals have no interest whatsoever in our needs or desires. It's our deal, not theirs."

Spending time with these outdoorsmen, I was amazed at their willingness to endure great hardship in seeking jungle wildlife. Phillips backpacked for days at a time through the high bush, blazing his own trails and sleeping among biting insects, spiders, scorpions, and serpents. His heavy load included electronic tracking equipment as well as camping gear, food, and water. Rodríguez thought nothing of searching for birds during rough weather and at hours when I would prefer to sleep in. The seasonal contrasts in Belize can be dramatic, ranging from hot, tinder-dry conditions to periods in which the deluge of rain is nonstop for days on end. I felt like a wimp, though I was doing my best to follow their examples.

On a bright spring morning I drove south from the field station into the 130,000-acre Gallon Jug Estate, gated and guarded, awash in some of the thickest lowland tropical forest I had seen in twenty-one years of exploration. The British military happened to be on maneuvers for "jungle survival training." Men and women in camouflage uniforms emerged from troop carriers emblazoned with the Union Jack and were swallowed by foliage within seconds. Despite the presence of soldiers, it was obvious that at least some wildlife had no fear of being hunted. White-tailed deer and ocellated turkeys—meaty game animals that are otherwise scarce throughout their range—grazed in abundance along the roadside, where shadowy forest met bright sunshine.

At a tiny farming center, dominated by an airstrip and cattle pasture, I arrived at the then-home of Carolyn and Bruce Miller. The husband-and-wife Wildlife Conservation Society scientists had lived and worked in the forests of Belize for two decades. Theirs is said to be a vanishing breed, as more wildlife research becomes relegated to urban laboratories and offices instead of direct field observation. The Millers, expatriate Americans who loved their work, occupied a comfortable home perched atop a steep mound of limestone (geologists would call it a karst formation). The hill was covered by thick bush, a sharp contrast to the flat, grassy livestock paddocks rolling across fenced savanna below. The yard and pitched roof of the Millers' ranch-style house held antennae for television, shortwave radio, telephone, and the Internet. Among other duties, this array kept ham radio operator Bruce in touch with the National Hurricane Center in Miami, with which he swapped storm-tracking data during hurricane season.

Carolyn and Bruce extracted themselves from his-and-hers computer stations at opposite ends of the large office they shared. Model-thin, blonde, and bespectacled, Carolyn is a world-renowned jaguar researcher who occasionally receives field assistance from her husband, a specialist on bats who bears the sedate and slightly rumpled look of a contented academic. They exude the air of cozy compatibility one associates with couples who spend most of their waking hours together—and who, because of shared interests, adore and respect each other all the more.

A leisurely conversation ignited in the tree-shaded yard, continued in the Millers' SUV en route to nearby Chan Chich ecolodge, then blazed over lunch in the hotel's elegant restaurant. The resort complex is owned—

like everything for miles around—by Barry Bowen, purportedly the richest man in Belize. Bowen's assets span from beer and soft drink monopolies to shrimp farming, cattle ranching, and a coffee plantation. Wilderness preservation and hands-off support of field scientists have been sideline passions of the conservation-minded entrepreneur, in his early sixties at the time of my visit and up for a knighthood from Queen Elizabeth II. Bowen belongs to a small but important cadre of successful businessmen whose contributions to environmental preservation, while derided by some at opposite ends of the debate, attract researchers and ecotourists in equal measure.

Bowen's Chan Chich Lodge is plunked into the middle of what was a medium-sized city some seventeen hundred years ago. Today its well-heeled visitors check in, watch birds, search for jaguars, eat fine food, sip cocktails, and snooze on mahogany-frame beds in a plaza built by the ancient Maya. Newer buildings among the old temples are made almost entirely of hardwoods harvested sustainably from surrounding jungle. Sadly, Chan Chich is of limited interest to archaeologists, as treasure seekers systematically looted the site long ago.

While Carolyn and I tucked into our chicken tostadas and Bruce munched on a cheeseburger, the three of us compared notes. Bruce was seeking new funding for bat studies, and Carolyn was analyzing jaguar photos from camera traps. She was on a first-name basis with a dozen or more cats prowling Gallon Jug. Our exchange lurched briefly to a halt when a skink skittered by and the Millers simultaneously shouted out the animal's scientific name. "Haven't seen one of those in a while," crowed Carolyn, beaming with glee.

From the outset, both Millers were eager to outline their interest in the biological corridors used by large animals to get from one place to another. They theorized that some species may travel farther and more often than is generally assumed, suggesting that the spread of houses, roads, farms, and even drug traffickers may have unexpected impacts on wide-ranging predators. Emerging evidence suggests that even shy creatures, including jaguars, find ways to cross human-inhabited areas without being detected, probably in the dead of night. Though such transits represent a hopeful sign, little is known about them, or about how fragile they might be. Were barking dogs enough to discourage jaguars? Busy highways? Bright lights? Tall fences? Open fields? If wildlife corridors are to be protected, the Millers maintained, such questions need to be addressed.

"All animals require the opportunity for genetic interchange," Carolyn said. "If species can't do this, they risk isolation and inbreeding." The latter, in turn, can increase the incidence of birth defects, reproductive failure, and physical abnormalities.

"As a young animal matures," Carolyn continued, "there may be no room near its birthplace to live as an adult. Perhaps it has to jostle or even fight for living space. Maybe it replaces an elderly male, or a robust adult drives it away. Isolated parks, no matter how magnificent, are not enough without genetic enrichment" from outside their borders.

I learned that reproduction among jaguars is modest even under ideal conditions. Cubs must be guarded carefully to be kept safe, and they depend completely on their mothers for a considerable span of time. Offspring are tiny and blind at first, opening their eyes within two weeks. Suckling continues for five or six months, but the cubs do not attain their full size and sexual maturity for two to four years.

Our talk took a more ominous turn when the Millers confirmed that Gallon Jug Estate was up for sale, at about thirty million dollars, and its future as a refuge was uncertain. (I later learned that in the absence of a buyer, Bowen consented to ecofriendly logging and oil exploration in order to raise cash.) My dining companions noted that the sprawling Yalbac Ranch to the south, which had protected adjacent forested areas, also was for sale. The properties were too expensive for outright purchase by environmental groups, and Belizean politicians were reluctant to expand the nation's already large network of national parks. About 45 percent of the country's land enjoys some degree of official protection, and there was limited support among Belizeans to add new reserves. (In comparison, only an estimated 3 percent of jaguar habitat throughout the cat's range is protected.)

"My interest is in seeing this area preserved long-term," conceded Carolyn, leaning forward to drive home the point. "It's a wildlife factory,

Figure 10.3. The Belize-based jaguar researcher Carolyn Miller setting camera traps on Gallon Jug Estate, near the Guatemala border in northwestern Belize. Shutters are tripped by the movement and heat of passing animals. Photo by Carolyn M. Miller

well protected and buffered from settlement on all sides. The forest is in fine shape and has a really good prey base. In short, this is an excellent place for jaguars." Statistics bore her out. In 2004–2005, Carolyn conducted a survey that estimated Gallon Jug's jaguar population to be about eleven per thirty-nine square miles (one hundred square kilometers), one of the highest densities ever recorded anywhere. She believed that the area—with the adjacent Programme for Belize and Yalbac properties included—supported about one hundred jaguars, easily 10 to 20 percent of the nation's total population.

"The cats are very conspicuous here," said Carolyn, adding that she had seen thirty-six individual jaguars over the years. "Many visitors to Chan Chich Lodge see jaguars; there's an average of about one sighting per week. Any guest has a decent chance of seeing one, and there are always tracks." Also common were cheek-rubbed tree trunks and leaf-litter scrapes. According to Carolyn, the felids of Gallon Jug seemed to have established a workable equilibrium with one another—what biologists call a tenure system—based on divisions of time and space.

I asked about personality differences between specific cats and was told about one male that knocked cameras askew three times, pointing lenses at the sky instead of where he wished to stride. This cat seemed determined to avoid being photographed. Another sly jaguar refused to leave his footprints on a smooth tracking transect even though the surface was made of all-natural materials.

"Females as a rule seem to evade cameras," Carolyn noted, repeating a common observation in the research community. "There are comparatively few photos of female jaguars or their cubs. I don't know why." Jaguar mothers, Carolyn declared, are among the most secretive of all cats. Their survival skills are honed to an extremely high level in order to protect their young, a high percentage of which fail to survive to maturity. Yet even lactating female jaguars occasionally succumb to poachers' bullets, which portends certain death for their motherless young.

I asked the Millers whether they could confirm a report earlier in the week that an unnamed Mennonite had admitted to killing twelve jaguars in northern Belize. "It wouldn't surprise me," said Carolyn, recalling that a Mennonite hunter was allowed to shoot one of her study animals because the cat, which had lost an eye, had become a cattle killer. "But non-Mennonites hunt jaguars, too. We can't control hunting. It really is a difficult problem that will take the government and the Mennonites themselves to solve. But some things *can* be done: better animal husbandry, to start, and keeping livestock well fenced. And with less hunting, the cats' natural prey may return to the forest."

After our meeting I confirmed that a Mennonite man, whose name was not disclosed, had been cited by the Belize government for possession of twelve jaguar pelts, stashed in the hunter's freezer. The Forestry Department confiscated the pelts and assessed a minimal fine on the offender, who was said to be "politically well-connected." A local newspaper later

ran a series of investigative articles that detailed the depth of the country's
illegal fur trade. En route back to Belize City, I interviewed several Mennonite residents at the Blue Creek post office and mercantile. None would
comment on jaguar hunting directly, other than to insist that its merits were
vigorously debated within the nation's Mennonite community.

The attitude of many Belizeans toward wildlife, Carolyn Miller believed, was shortsighted: "I hear people saying, in effect: 'There it is; I'm
going to shoot it.' The law clearly states that you are not supposed to hunt
jaguars, yet people do." The situation presents a quandary, she explained.
"Conservation is being asked to address so many sociopolitical problems.
What we really need is effective environmental education of the next generation." Carolyn echoed a theme I heard often: the conviction that new
attitudes and effective solutions will come from young people, not from
their parents and grandparents. The gist of Carolyn's argument was that
without a willingness to break away from tradition-bound thinking among
all parties, including conservationists, creative solutions to poaching and
indiscriminate habitat destruction would remain elusive.

Bruce Miller chimed in to suggest that Belizean tax policies made it
difficult for conservation groups to buy and hold substantial acreage for
wildlife protection. "The government virtually demands that you develop
large pieces of land," he said, "rather than let it sit idle. Plus, a minister can
'de-reserve' any park simply by fiat. Protection is not permanent. Another
problem is that many immigrants simply peel the forest off when they move
to a new piece of land, exhausting its resources within a few years." Belize,
like many Latin American countries, had actively encouraged homesteading
for decades, and this policy was inexorably squeezing animals into smaller
patches of land.

We discussed the pros and cons of relocating large carnivores away
from areas where they came into conflict with transient immigrants or local residents. Bruce felt that physically shifting wildlife solved nothing. He
cited 2005's Hurricane Katrina disaster as supporting evidence, even though
the "animals" involved were humans: "When New Orleans residents were
moved posthurricane into communities already occupied, people often did

Figure 10.4. This Belize jaguar, nicknamed Fenton by the wildlife biologist Carolyn Miller, was tracked and shot in 2005 after killing a calf. Miller had observed this adult male for several years before it was declared a nuisance. Photo by Carolyn M. Miller

not adjust well. In many cases, the cities into which storm victims moved had high population densities. Similarly, in nature, when a habitat is destroyed or radically changed, animals may be forced into places where others of their kind already live. Related social issues for cats are poorly known."

One underlying—and potentially intractable—problem with relocating jaguars is that cats seem to possess a distinct homing sense. This may be the result of a neurochemical response to solar positioning in their native geographic location. In his book *A Cat Is Watching,* the naturalist Roger B. Caras posited that a felid "has a sun compass. If you move it 200 miles you throw the angle of the sun off by a significant degree. A cat sensitive to that change would almost certainly move in the direction that tended to correct the change, i.e., toward familiar home ground." The experience of wildlife managers (and pet owners) bears this out; even relocated housecats sometimes return in short order to "home" addresses miles away.

The Millers described documented instances in which jaguars forcibly moved a distance of many miles were back "home" within days. Similar behavior has been recorded among big cats in other countries. Mountain lions in the United States, for example, have made documented journeys of two hundred miles or more to their original territories after relocation by truck.

I bid the Millers adieu and journeyed back to La Milpa Field Station, my head buzzing with new information. I was excited to be driving through the heart of jaguar country. Carolyn had confirmed claims made by Vladimir Rodríguez and Ryan Phillips that local cats often traveled on local roads, so I stayed vigilant as I headed north. The forest arched over my truck, turning daytime into dusk. I felt hopeful that Carolyn's "wildlife factory" would somehow be preserved, despite encroachment from many quarters.

Two days later, after foot searches in every direction from La Milpa, I broke through the green tunnel at the Rio Bravo property line and crossed again into treeless pastures saturated by the white glare of unobstructed sun. I stopped, as instructed, at the first house on my left. I parked in front of a machine shop and startled a fresh-faced girl wearing an ankle-length plaid dress and a broad-brimmed bonnet. A boy at her side wore the standard male Mennonite uniform of dark pants, blue shirt, black suspenders, boots, and straw hat. The pair seemed to have trouble understanding English. They pointed to a man in the workshop, who cocked his head wordlessly in the direction of a stairway attached to the farmhouse.

The cramped upstairs nook was adjacent to a high-ceilinged attic where a teenager with cornflower eyes, milk-white skin, and rosy cheeks folded laundry. She smiled pleasantly and gestured toward a monitor and computer, from which a coaxial cable trailed out the window to a satellite dish. A jelly jar was set out to collect the Internet access fee of one dollar per fifteen minutes. I logged into cyberspace, light years away from this uneasy tapestry of farm and forest.

CHAPTER ELEVEN

"Cows Are More Important Than Cats"

CARLOS LÓPEZ GONZÁLEZ, accompanied by a north-of-the-border gringo, sat in sweltering heat, swapping tales with the head of a poor *mejicano* family in the Mexican state of Sonora. As recounted in a magazine article co-written with a colleague, David E. Brown, the Mexican researcher listened patiently to an elaborate description of their informant's encounter with a jaguar.

"Whenever [the hunter's] account lagged for even the briefest moment," the wildlife biologists wrote, "his wife or daughter jumped in with the missing details, retelling the story they had probably heard a hundred times. Killing the *tigre* was likely the highlight of his life, and therefore of theirs. Enthused about his recollections and encouraged by the rapt attention of his audience, the man let his wariness subside. Suddenly he wheeled around and disappeared into the bedroom." The hunter returned with skulls of a jaguar and a mountain lion, as well as the skin of what the man believed to be a jaguar cub.

After López González and Brown correctly identified the pelt as that of an ocelot and distinguished between the jaguar skull and that of a mountain lion, the man let down his guard. "We must have passed some sort of test," the visitors recalled, "because soon we were hearing of other jaguar and ocelot hides in other villages. The door had opened."

This recounting might not surprise many south-of-the-border scientists. "Jaguars are intertwined and enmeshed in thirty centuries of Mexican culture," said Roberto Águilar, a wildlife veterinarian of Mexican descent and director of conservation and science for the Phoenix Zoo. "When you speak of jaguars in Mexico, it brings up a sense of deep history, pride, and empathy for an animal that has permeated [the national] culture."

There also lingers, Águilar told me, "a sense of fear and competition" toward jaguars. "But in general Mexicans are proud of jaguars and want to keep them around for generations to come."

López González, a University of Querétaro professor, and Brown, of Arizona State University, had been meeting with jaguar hunters in a rugged area about 130 miles south of the Arizona border town of Douglas. This isolated region of Sonora, in the hardscrabble Sierra Madre Occidental, is dotted with scrubby, arid woodland inhabited by deer, coatis, and peccaries—and the jaguars that eat them. A breeding population of *Panthera onca* has persisted around the farm and ranch villages of Huásabas, Nácori Chico, and Sahuaripa, near the confluence of the free-flowing Aros, Bavispe, and Yaqui rivers. Many experts believe that this is the likely source of the jaguars seen in the United States since 1996. The borderlands of Arizona and New Mexico are within the known travel range of dispersing young males. Several uninhabited mountain ranges form convenient corridors across the international boundary.

The professors were in Mexico to search for jaguars under sponsorship from several groups, including CNN founder Ted Turner's environmental research foundation and New York's Wildlife Conservation Society. The men found what they sought, but their discoveries confirmed an alarming trend: an estimated 30 jaguars had been killed in northeastern Sonora during the preceding decade, including at least 6 animals in 1999 alone. Of these last, 4 had been cubs or lactating females. At that rate, with no more than 150 cats believed to remain in the area and travel routes of more southerly jaguars in jeopardy, it was feared that the population could disappear before long. Enduring cultural traditions that supported such indiscriminate killing made extirpation appear even more likely.

Hunting exotic cats is a source of great personal pride for many men—and perhaps some women—in countries throughout the jaguar's range, including the Spanish-speaking nations of Latin America. Some aspects of this may reflect the cultural trait known as machismo.

Defining machismo—or, in the Anglicized rendering of the term, "being macho"—is tricky. In literal translation from Spanish, machismo is simply "being male" or "manliness." But in modern English usage the term refers to projecting prominent or excessive masculinity in attitude and behavior. This may be interpreted in a positive way: as being courageous, strong, aggressive, or virile. In Latin America, machismo is frequently not only accepted but encouraged and admired. Men are often assumed to have a sense of entitlement and the right to dominate. Among non-Latin Westerners, however, calling someone "macho" or referring to his "machismo" may be derogatory, an implication that the man so labeled is overly controlling, inflexible, abusive, pompous, and even violent toward those who are weaker, including women and children. Such a man may be viewed as a boor, philanderer, fighter, or hell raiser, cartoonish in his exaggerated masculinity.

When a "macho" man—of whatever nationality—decides that killing a jaguar will enhance his masculine virtues, the hombre's desire may outweigh such niggling details as legality, expense, and obligations to family, friends, church, or employer. Such a fellow may be willing to spend thousands of dollars and put his life or job at risk for a shot at a jaguar, knowing full well that he could be fined and thrown in jail—or forced to pay a bribe in order to silence authorities.

The Brazilian field biologist Sandra Cavalcanti, who has studied jaguars extensively in the wildlife-rich Pantanal district, has confirmed that among the area's residents an ingrained belief prevails that "killing a jaguar is considered a manly thing to do." Given the high rate of poaching by *los machos* and others who may be oblivious to its long-term consequences, Cavalcanti and a growing number of her colleagues have concluded that simply studying jaguars is no longer sufficient. They believe that neutral scientific observation of fast-disappearing animals without parallel action to defend them is a luxury researchers can no longer afford.

Machismo may be among *Panthera onca*'s deadliest enemies. Despite restrictive laws virtually everywhere the jaguar exists, there seem to be plenty of individuals who feel compelled to track and kill this cat even as the animal vanishes. Even those in the highest ranks of officialdom may be tempted. How else can one explain, for example, a 1997 decision by the Venezuelan government to permit the legal hunting of jaguars, allowing as many as thirty of the cats to be exported as trophies each year? This proposal was advanced by a nation where killing jaguars has been against the law for decades and an estimated eighty to one hundred are still poached annually.

Convictions, especially in rural areas, have strong roots. "Hunting [traditionally has been] the passion of the men of the plains, the measure of their strength and cunning," noted the journalist Parisina Malatesta, in a magazine description of the mid-twentieth-century Venezuelan *llanos* or savannas. "In summer, people came from all over the llanos to hunt jaguar."

Apparently unable to control illegal harvesting, Venezuela's wildlife authorities were eager to collect the $10,000 to $15,000 hunters would pay for permission to kill a jaguar in its natural environment. Hunting guides and outfitters also could make money on the deal. Responding to widespread outrage—and the likelihood that the Convention on International Trade in Endangered Species would reject its request for a waiver from the ban on jaguar trade—Venezuela withdrew its proposal. But the fact that it was considered underscores the enduring thrill of "jaguar sport."

What is the underlying appeal of such recreation? While I have respected friends and family members who hunt, the pastime has never drawn me in. Certainly there is great skill and excitement involved in tracking and subduing game, and we are indebted to hunters for helping us with the necessary management of wildlife. But why is the jaguar, in particular, a favorite target, beyond its qualities as a formidable predator?

I found a few clues in a travelogue entitled *Lost Trails, Lost Cities,* by Colonel P. H. Fawcett. A lifelong British adventurer, Fawcett was an explorer in the grand colonial tradition. He tramped through the unmapped heart of the Amazon Basin between 1906 and 1925, sending newspaper dispatches that captivated readers in the United States and Europe. A scientist and engineer by training, Fawcett was sent to Amazonia in order to determine the exact boundary between Peru and Bolivia. He parlayed that assignment into an excuse to experience all that the frontier offered. Fawcett wrote the following between 1906 and 1909, while based in Rurrenabaque, Bolivia: "Jaguars are very common on the cattle plains, and the great sport is not to shoot them but to lasso them on horseback. Two men take part, keeping the roped beast between them. It requires good mounts and considerable dexterity with the lasso, but given these it is not nearly so dangerous a sport as it sounds."

An avid collector of colorful anecdotes, Fawcett made the claim that "jaguars can sometimes be tamed, and do not make dangerous pets if caught as cubs. There was a practical joker at Reyes, a few leagues from Rurrenabaque, who had quite a large [jaguar] that he allowed to wander like a dog about the house. His great delight was to take his pet along the trail toward Rurrenabaque and wait for travelers on muleback. At a signal the jaguar would leap out from the bushes and the mule would bolt, usually unseating its rider, whose terror at finding himself face to face with the beast can be imagined."

Fawcett disappeared in the Brazilian jungle in 1925. Some claim to have discovered his bones and believe he was killed and eaten by tribal people. Conclusive proof never surfaced.

While the days of such unfettered adventuring may be over, hunter and jaguar still engage in their deadly tango. I heard jaguar-hunting stories in every country I visited, from Mexico to Panama. (My travels in South America were limited to Ecuador.) In Belize, for instance, I heard reports of foreigners offering villagers $10,000 to $15,000 for the fresh pelt of a jaguar. "The Belize government," an informant told me, "confiscated thirteen skins from a single dealer, then searched and found another seventeen carcasses in the man's possession." A separate report claimed that a single rancher had killed twenty-two jaguars within a three-year period. What he did with the pelts is unknown. Farther south, in coastal Honduras, I was told that the Garifuna people—whose mixed-race ancestors are from the Carib and

Arawak tribes of the Caribbean as well as from West Africa—continue to harvest jaguars because they traditionally stretch the cats' skins over the heads of their ceremonial drums. Traditional Garifuna reportedly use meat and fat from these carcasses in protective medicines and ointments.

Still farther south, some rural Brazilians still prize *carne de onça*—jaguar meat—and consume it when available. The flesh, described in a 2007 news account as "delicious," is reportedly distributed as a bribe to game wardens and other authorities when poachers and ranchers find such payoffs useful. The remaining jaguar meat is apparently sold on the black market.

In Ecuador during a monthlong visit, I shared a beer with a self-described "unrepentant" poacher who claimed to have shot and skinned more than four dozen *tigres* in the country's tropical lowlands.

"The settlers who live there hate them," declared the grizzled hunter, who proceeded to describe jaguars as "worth much more dead than alive. They were happy to hire me to track and kill the varmints, and I probably would have done it even if it didn't pay good money."

The sixtyish man of indeterminate nationality paused to sip his *cerveza*. He grasped the frosty mug with a left hand that seemed to have lost an index finger to bush knife or machete. "A critter that smart," he continued, "is so much damn fun to hunt, and I'd still be doin' it except for one thing."

"What's that?" I asked.

"Ain't no more out there," he shrugged, waving his mutilated hand in the general direction of the upper Amazon. "At least not enough to matter."

Sonora's Sierra Madre Occidental was a destination well known to sport-hunters during the 1930s, when the Arizona guides Clell and Dale Lee led backcountry expeditions yielding eight jaguars over a three-year period. A dwindling number of breeding jaguars are still there, scattered along the western edge of the Continental Divide. At least two smaller groups of jaguars remain elsewhere in Sonora: one in its southwestern Sierra Bacatete, the other along the border with Sinaloa. There may be more cats scattered around, but their numbers are exceedingly small.

Northeast Sonora's jaguar country is the three thousand–square–mile home to a unique mix of flora and fauna. The steep, parallel ranges are high enough to catch stray moisture, creating a host of microclimates. Native vegetation ranges from cacti, mesquite, and palm trees at lower levels to ponderosa and piñon pine trees in the upper elevations. Tropical thorn-scrub (a specific type of vegetation characterized by spiny, small-leaf trees, shrubs, and succulents) is also evident, along with deciduous forests that include oak and sycamore. In this domain one finds the southernmost bald eagles and northernmost military macaws, along with boa constrictors, Gila monsters, neotropical river otters, and nectar-loving bats. Scores of species of butterfly—some new to science—flit among the slopes and peaks, along with lilac-crowned parrots, elegant trogons, and vermilion flycatchers. Two of the Sierra's largest carnivores, the grizzly bear and Mexican wolf, once well established here, were extirpated by the mid-twentieth century.

Long dominated by hacienda-style cattle ranching, this is a land of free-flowing rivers, fallow meadows, and uncut forests. One can stand on a ridge and view hundreds of square miles unbroken by road or fence. Animals wild and domestic quickly vanish into the undulating landscape. Resident jaguars roam almost everywhere, but seem to prefer obscure, tree-shaded canyons and brushy thickets, feeding mainly upon wild mammals and the occasional stray calf.

A logical question emerges: with so much indigenous prey and little-disturbed habitat, why is long-term survival of Sonoran jaguars threatened at all? The biologists López González and Brown cite economics as the chief culprit. "The reason jaguars are still being killed in Sonora," they contend, "is that cows are more important to Sonorans than cats." Older ranchers in particular despise top predators and will poison, fatally trap, or track them with hounds. The reality, according to these scientists, is that "cattle range everywhere [in the Sierra Madre] and have become a regular part of the jaguar diet." What portion of that diet is unknown.

Rural livestock management is at issue here. In much of Latin America, including Mexico, husbandry has changed little since Spanish missionaries and settlers introduced domestic animals to mission communities in the sixteenth century. Cattle often wander freely and unprotected, reproducing at will and becoming virtually wild. A pertinent contrast with U.S. practices is instructive. Cold winters north of the border require calendar-scheduled breeding. Calves are born in protective pens or barnyards during spring. This allows a calf to come into the world after snows have melted, then acquire strength and savvy through summer and fall.

In contrast, Sonora's rugged terrain is nearly roadless, making it difficult for ranchers to manage cattle carefully. Free-range bulls sire whenever and wherever they please, giving felids and coyotes a chance to feed on newborns or young calves. Restricting mother cows to enclosed pastures and fencing cattle out of isolated terrain would help reduce the depredation, experts agree, but it is impossible to alter topography and not easy to change five hundred–year–old habits and attitudes.

Environmental education—or lack of it—is also part of the mix. Many who indiscriminately kill cats in Mexico (and other countries) have little formal schooling and limited media access. They may be unaware that jaguars are endangered animals fully protected by both national and international laws. Or they may believe unfounded rumors and folklore suggesting—incorrectly—that jaguars are innately aggressive toward people. Also, many in ranch country regard *all* jaguars as unwelcome predators, whether or not an individual cat takes livestock. (According to researchers, most jaguars and mountain lions do not become cattle killers unless the cats are sick, injured, or weak, or for other reasons cannot hunt wild game successfully.)

"In the end," Juan Carlos Bravo, who represents the Mexican environmental group Naturalia in Sonora, told an interviewer, "large-carnivore conservation has a lot more to do with politics, economics, and society than with biology. . . . We can do all the field research we want and it won't be enough if we don't have state governments, local ranchers, and the general public in favor of preserving jaguars."

Bravo believes that most Mexicans, particularly city dwellers, are unaware that jaguars continue to roam their country. In contrast, rural residents who do find the cats on their properties often expect help in lowering depredation losses as a prerequisite for support of jaguar conservation.

"We visited a large cattle ranch whose owner was paying three of his cowboys to shoot jaguars on sight," the Mexican biologist Sergio Ávila told me, during an interview with Emil McCain, his then-partner in Sonoran big-cat research. According to Ávila, "The rancher paid each cowboy a bounty equivalent to triple his monthly salary whenever that *vaquero* killed a jaguar." Workers on this ranch had shot several jaguars during the previous few years, along with mountain lions, coyotes, and even turkey vultures (which feed only on dead animals).

"It didn't matter whether any of the cats were destroying the man's property," McCain explained. "This rancher simply didn't want jaguars around."

The scientists were infuriated, particularly since the ranch owner lived in the Sonoran capital of Hermosillo and had little direct experience with wildlife. The researchers took matters into their own hands by meeting with ranch hands and explaining the purpose of their big-cat study. Convinced of the worthiness of the enterprise, some vaqueros agreed to stop their wanton killing. Others signed on after Ávila and McCain offered to more than match—out of their own pockets—the local bounty paid for dead jaguars.

"It's all about economics for these people," Ávila said. "They are trying to survive under very difficult circumstances. Each day they get up and have to take some kind of direct action simply to feed themselves. You have to give them a financial justification for *not* killing a jaguar, because otherwise they will."

Octavio Rosas-Rosas of New Mexico State University in Las Cruces was the fifth jaguar researcher—after López González, Brown, McCain, and Ávila—in as many years to journey into the Sierra Madre Occidental. Like his predecessors, the baby-faced graduate student staked out a specific area in which to set up his camera traps and conduct ecological transects. (A transect is a systematic, ground-based census of flora and fauna on a measured piece of land.) But Rosas-Rosas wanted to do more. Like Ávila and McCain, he sought to broker deals with members of the community in hopes of persuading ranchers to end their persecution of *Panthera onca.*

"What I offer is a mechanism whereby ranchers receive 'alternative compensation' for any stock killed by jaguars," Rosas-Rosas declared at a 2004 meeting of the Jaguar Conservation Team, in Animas, New Mexico. This was accomplished, he explained, by paying people to leave jaguars alone—but with a twist.

In cooperation with Sonoran ranchers and business partners on the U.S. side of the border, Rosas-Rosas set up an innovative program whereby sport hunters from the United States paid a premium price for deer hunting on ninety-three thousand acres of private land owned by a dozen participating Mexican landowners. An outfitter brought U.S. hunters onto the ranches in order to shoot Coues deer, a white-tailed subspecies found only in the American Southwest and northwestern Mexico.

Rosas-Rosas introduced several Sonoran ranch managers at the Jag Team meeting. Through spokesman Jesús Moreno, they confirmed their commitment "to live with jaguars through economic diversification." In addition to receiving payment for use of their land, the ranchers hired out their homegrown cowboys as guides. Researchers, in return, were granted permission to study jaguars on private acreage and given shares of a conservation fee paid by the visiting hunters. Another stream of income flowed from ecotourists, who paid for backcountry excursions that offered a chance to glimpse a wild jaguar.

The biologist's project was well supported. He explained that he was working jointly with SEMARNAT, Mexico's Environment and Natural Resources Secretariat—roughly equivalent to the U.S. Department of the Interior—as well as the University of the Sierra Madre, the State of Sonora, the village council of Nácori Chico, and several nonprofit groups. A foundation funded by the clothing designer Liz Claiborne and her husband had pitched in, along with the Buddhist nation of Bhutan in the far-off Himalayas. This diverse coalition promoted education about jaguars in a fun and aggressive campaign that targeted children as well as adults.

Figure 11.1. A Camtrakker® image of Abelardo, one of several adult jaguars photo-graphed (and named) by Octavio Rosas-Rosas in the mountains of northeastern Sonora, Mexico, February 25, 2005. Many experts believe that such animals are the source population for U.S. jaguars. Photo courtesy of and © Octavio C. Rosas-Rosas

"I would be very happy to have any of you come to see what we're doing," Rosas-Rosas told his audience at the Animas High School library. After the meeting I introduced myself to Rosas-Rosas, a dark-haired fellow in his late twenties. I told him I wanted to be part of his winter-spring research season.

I made good on my promise seven months later when I flew to Tucson, rented a car, and sped three hours to the border. I checked into a Douglas motel, located next door to a maze of Border Patrol holding cells. The Motel 6 clerk asked whether I was a journalist in town to cover "the vigilantes." Unknown to me, scores of angry U.S. citizens were assembling for the pur-pose of spotting and apprehending anyone crossing the desert illegally from Mexico. When I assured the young man that I was covering a different kind of surreptitious entry—by Mexican jaguars—the clerk looked bewildered: "Wild jaguars! Around here? You're kidding."

I assured him I was not, and displayed copies of pictures Jack Childs and Warner Glenn had taken in 1996. "I think our vigilantes are much more dangerous than these cats," the clerk sniggered.

At the Gadsden Hotel the following morning I met Rosas-Rosas's partner, Gordon Whiting, an avid hunter, wealthy rancher, and former member of Arizona's Game Commission. Accompanying him from Phoenix were the big-game hunter Bob Walker and the businessman Melvin Bushman. After border formalities our journey took us in a pair of pickups from Agua Prieta south through open range toward Nácori Chico, ground zero for Sonoran jaguar research. We drove nearly five hours south on paved highways, then snaked east on sinuous mountain roads before settling for the night at a colonial-style motel in Granados. The serpentine Río Granados, a favorite destination for jaguar hunters before 1970, flowed unimpeded along the outskirts of this bucolic town.

The next morning we climbed from the fertile river valley into the heart of the Sierra Madre, passing sheer four hundred–foot cliffs where falcons and hawks were teaching their fledglings to fly on gyrating morning thermals. Soon the hardscrabble landscape became utterly devoid of people, and we entered territory dominated by stunted trees, spiky bushes, and lichen-etched boulders. Centuries-old walls of stone tumbled and zig-zagged across overgrazed slopes. At a moister elevation, we entered a peculiar world where sturdy evergreen oaks grew next to frilly fan palms. Prickly pear, organ pipe, and cholla cacti shared the environment with streamside willows, ferns, and cottonwoods. Well-hidden scorpions, Gila monsters, and rattlesnakes skittered and slithered among the rocks. Floral oddities included the lipstick-tipped ocotillo, the ghostly palo blanco, and the aptly named rhino tree. This last I recognized as a cousin of Central America's stately kapok, sacred to the Maya.

An hour east of Granados we left crumbling asphalt and turned onto one of the single-lane washboard tracks typical of the Sierra. These crude roads are the only overland routes other than backcountry trails to hamlets founded by the Spanish in the 1600s, still remote outposts of civilization. Residents rely almost entirely on subsistence farming and ranching. Many are desperately poor, and some, I was told, accept payoffs from smugglers moving illegal drugs or undocumented immigrants toward *el norte.*

It took more than three hours to cover the forty-seven miles to Nácori Chico, a village built in a defensive manner to ward off once-frequent attacks. Its houses face a central plaza dominated by a stately whitewashed church. Adobe walls are linked, originally intended to prevent outsiders from forcing their way into town. Chiricahua Apaches controlled the nearby spine of the Continental Divide through the nineteenth century, and, remarkably, a few managed to cling to their traditional way of life in remote canyons well into the 1930s.

We loaded last-minute supplies at a "mini-super" and drove on even rougher roads—impassable in bad weather—into a mountain range appropriately called Los Tigres. All the land around us was privately owned, much of it reportedly used to bring marijuana, cocaine, and heroin along the same paths trod by spotted cats. During my visit I saw an unmarked airstrip and several mule trails said to be used by *narcotraficantes.* Rural travel here was not recommended without an armed escort.

At midday we reached the humble headquarters of Rancho Napopa, one of several extensive cattle operations in the region. Such spreads typically encompass ten thousand or more acres, with livestock herded by cowboys on horseback in the time-honored manner. Napopa's owner, Jesús Moreno, was the man I had heard speak in New Mexico about the Rosas-Rosas project. Moreno's headquarters consisted of a simple cement-block house, several livestock corrals, and a few outbuildings. Amenities included a well, a stock pond, solar-powered lights, satellite telephone, a flush toilet, and an indoor shower.

A fresh jaguar track had been found earlier in the day only a half-mile east of the *rancho,* and after lunch we set out to find it. Guiding us was the U.S. researcher Dave Strozdas, whose principal occupation was studying mountain lions on one of Ted Turner's properties in New Mexico. The lions were systematically killing off the Turner ranch's desert bighorn sheep, a species struggling to survive in the region. A broad-chested former Oklahoma game warden with a lawman's friendly but firm demeanor, Strozdas helped Whiting and Rosas-Rosas set up hunting and ecotourism tours of the region.

We were led through thornscrub to a dusty section of trail that, following a recent rainstorm, had captured a fist-sized track. Strozdas removed a protective shield of rocks from the path to reveal a perfect impression of the four circular toe pads and wide plantar (heel pad) of an adult jaguar. An expert can easily identify the differences between the tracks of a jaguar and a mountain lion, but distinctions are subtle to untrained eyes. There are variations in toe pad length and width, for example, and most aspects of the jaguar track are rounder than those of a mountain lion, with less pronounced lobes and creases on the sole.

"Look at the size of that thing," whistled Bob, the trophy hunter in our group. "That's the biggest cat spoor I've ever seen."

Dave and Melvin, the Phoenix executive, fashioned a simple frame and poured yellow dental plaster to create a three-dimensional copy of the mark. Leaving the plaster to set, we hiked to a second jaguar track cast earlier in the week. The trail immediately turned rocky and steep, choked with nearly impassable knots of woody vegetation. At a high point, we scanned hundreds of square miles of blue-green mountains. The only obvious indicators of human presence were herds of cattle and sheep that browsed a nearby hillside. A trained dog guarded the sheep, but the cows ranged unattended.

The second set of jaguar tracks traversed the saddle of a ridge where two trails intersected. We all laughed upon seeing that the prints were embedded in a phonebook-thick pile of cow dung that had hardened to rocklike consistency. After photographing the impressions, we scrambled down a precarious slope to a pinch point where a camera trap was concealed. There were twenty-six such setups in Rosas-Rosas's study area, each holding a pair of camouflaged cameras on opposite sides of game paths. The presence of Sonoran jaguars was well documented here, along with mountain lions, bobcats, ocelots, deer, peccaries, and whatever else wandered by.

We returned from our trek, ate a simple dinner, and relaxed with

Figure 11.2. Mexican college students conduct a transect of known jaguar habitat in the Sierra Madre Occidental during spring 2005. Such precise studies of a designated area can suggest the number and variety of prey animals available to large cats. Photo © Richard Mahler

Rosas-Rosas and the eleven Mexican university students he had invited to spend Easter vacation conducting field surveys. Rosas-Rosas asked me to interview each student about his or her experiences. I learned that half were from a newly built community college in Moctezuma, a small city in northern Sonora, while the others attended an exclusive private university in Puebla, Octavio's hometown.

The students' passion for conservation and devotion to fieldwork surprised me. Earlier that day a group had discovered, after extensive searching, the newest set of jaguar tracks. Each participant called this discovery a highlight. Several expressed great pride that jaguars were still found in Mexico. Most had never been to such a remote outback, or even into the countryside. During our conversations, a few students said they thought it better to preserve jaguar habitat by keeping it in private hands and controlling visitation than by creating a national park or sanctuary, as would be likely in the United States. "In Mexico," María told me, "property rights are almost sacrosanct. You try never to separate people from their land." Others nodded in agreement as María summed up: "In Spanish, the word *pueblo* means both 'the people' and 'the land'; so who we *are* also is where we *live*."

At sunrise I joined Rosas-Rosas and his students on a transecting expedition to a neighboring property, Rancho los Pescados. Our four-wheel-drive pickups crept through a steep-sided canyon that at first appeared impassable. Much of the route did not follow a road at all, but instead navigated bumpy cobbles at the center of a flowing stream. Later our trucks clung at sideways angles to a nearly vertical horse track. At one point we narrowly avoided tumbling into a ravine severely eroded by heavy rains. (The Sierra had been blessed with its greatest winter moisture in fifty-six years.)

After an hour of slow travel we entered a broad valley, watered by a tree-choked creek and exploding with fresh greenery. Rancho los Pescados had been cattle free for eight years, and its undulating landscape was thick with knee-high grass and bright with wildflowers. The ranch owner, an absentee, was said to be in financial trouble. The man had obtained a conservation easement whereby he was paid by the government for allowing his land to return to a seminatural state. We saw ample proof of its rejuvenation, including several Coues deer peering from the underbrush.

At a severe washout we were forced to stop and park. A meadow, blazing gold with Mexican poppies, provided the students with a good transect location. This involved marking off a specified distance from a starting

Figure 11.3. The Mexican biologist Octavio Rosas-Rosas checking a Camtrakker® camera trap strapped to a tree trunk in the Sierra Madre Occidental of Sonora, circa March 2005. Photo © Richard Mahler

point and examining intervening ground for any and all evidence of animal tracks, scat (feces), scrapes, sprays, hair, and other biological signs. Over the next two hours the group found physical evidence of peccaries, deer, coyotes, bobcats, skunks, and mountain lions. At the far end of the transect we followed a dry creek onto a game pathway squeezed between a copse of oaks and a rocky outcrop. A camera trap checked here had captured only fourteen new images, so it was left unopened.

Although we saw no jaguar tracks on this outing, Rosas-Rosas was excited to see plenty of food sources. He speculated that an increase in prey density was caused partly by ample rains, which had increased plant growth, as well as by the continued absence of hungry livestock. "This might explain a decline in livestock depredation by jaguars," said Rosas-Rosas. "No cows have been killed by a jaguar near here for the past eight months. In drought years more than twenty head might be taken in a single month."

During the two-hour journey back to Napopa, a mounted cowboy appeared from out of nowhere, tipping his hat in a courtly gesture. The vaqueros typically received no more than $150 a month, plus room and board, for their difficult work. The men are separated from their families for weeks on end and must patrol harsh country under a relentless sun. Some now made extra money helping jaguar researchers. That evening, as the full moon rose, several cowboys came to the house and expressed gratitude for our gifts of whiskey, song, and companionship.

"You've given us good reasons not to kill jaguars," confided one of the men in slurred Spanish, after delivering a stirring rendition of Mexico's classic tear-jerking ballad, "Volver Volver." "They are very brave and strong fighters. *Que viva el tigre* [Long live the jaguar]!" I shook the fellow's weathered hand and encouraged him to keep Sonora a safe harbor for the big cat. He gave me a sly wink and launched into a song made famous by Los Tigres del Norte, a phenomenally popular Mexican band whose members sometimes wear fancy suits that resemble jaguar pelts.

The following day Rosas-Rosas beamed upon seeing a large new billboard that loomed at the entrance to Nácori Chico. Dominated by a full-color image of *Panthera onca*, it welcomed visitors to "the land of the jaguar." Rosas-Rosas proceeded to tell me that area schoolchildren had held a contest to design a T-shirt promoting jaguar conservation, and that a Spanish-language cartoon was about to debut on nationwide Mexican TV starring "Mr. Jaggy," a superhero based on Sonoran cats. Building on this goodwill,

Rosas-Rosas said he hoped to expand from twelve to nineteen the number of participating ranches in his study zone.

Between 1999 and 2007, Rosas-Rosas identified seven individual jaguars—including three females and two cubs—north of the Río Aros. He named one young male Panchito, after the Mexican revolutionary Pancho Villa, who invaded the United States briefly in 1916 from Chihuahua. Other favorite jaguars of Octavio were dubbed Juanita and Abelardo. Before leaving, he gave me a grainy photograph of the latter cat, whose eyes glowed like embers in the camera's bright flash.

"We tried very hard to trap and collar at least one," Rosas-Rosas allowed, "but had no luck. It was like they were trying to outwit us. I'm optimistic that we can save the Sonoran jaguar, but it will take all of us working together for this to happen."

CHAPTER TWELVE

"We Just Stopped Seeing Them"

I FIRST MET SERGIO ÁVILA and Emil McCain in Tucson, where the researchers were studying a pair of captive sister jaguars at a local biological park. "We've been annoying the zookeeper," McCain laughed, "by getting in the [unoccupied] cage to have a look at jaguar tracks."

On a seasonally hot afternoon, we dined beneath a web of misting tubes at a restaurant near the University of Arizona. In collaboration with Jack Childs—whose party videotaped the Baboquívari jaguar in 1996—McCain was monitoring camera traps along Arizona's southern border, with occasional assistance from Ávila. The Mexican biologist's research background primarily involved mountain lions of his home country, while McCain had done fieldwork related to various felids in the course of graduate school studies. Both were energetic young men, trim and tan from long days outdoors.

McCain, with his close-cropped red hair and beard a dead ringer for the Dutch painter Vincent van Gogh, opened the three-way interview by speculating that jaguars "may be the hardest of all large cats to study in terms of habit and habitat. They live in places that are hard for humans to get to and to work in, but also the animals themselves are largely nocturnal, extremely quiet, and notoriously secretive." Part of their elusiveness, he explained, is hard-wired behavior. Jaguars, like most cats, often remain hidden in order to get close to their prey and attack from a short distance. For such ambush predators, staying unseen is an essential skill.

McCain might have added that the jaguar's renowned shyness also makes it expensive and time-consuming to study. In six months of tough Sonora fieldwork during 2003, he and Ávila had used snares to catch only two cats, neither monitored long enough to provide much data. The two men had snared one female jaguar in Sonora that slipped its collar, while a second jaguar, an adult male, died during the capture process. "I am haunted every day by that loss," confessed McCain, citing a host of factors contributing to the latter cat's death, including the region's intense, dry heat.

Octavio Rosas-Rosas had tried even longer and failed to capture a single jaguar. I later learned that such experiences were not unusual: some researchers have spent years studying jaguars in the wild without ever glimpsing one.

"Trapping and collaring a jaguar successfully is very difficult," explained the fast-talking Ávila, who wore his black hair in a thick ponytail. Collaring, he said, allows researchers to follow a cat's movements with precision, providing clues about its hunting, sleeping, and mating habits as well as its travel patterns.

Snaring a jaguar is a delicate operation that essentially involves setting a trap in which the cat becomes entangled in a web of knotted rope. Felids are suspicious and cautious by nature, which makes them difficult to trap in the best of circumstances. Jaguars—which do not sweat—overheat easily in capture situations and may fatally injure themselves while trying to escape. Such commonly used tranquilizers as ketamine must be measured and injected carefully so as not to under- or oversedate. Weight is only one factor in estimating dosage. For unknown reasons, some cats are more sensitive to tranquilizers than others, and this tendency has been noted among northern jaguars. Despite such hazards, both scientists felt that radio-collaring, where appropriate, was an invaluable research tool.

"Location is only one part of the data you collect," said Ávila. "You want to know what they're doing, how long they stay in a place, what the weather is like, whether they're alone, and so on." Without ongoing tracking of a collared jaguar, much of what might be attributed to such a free-ranging cat is guesswork.

Although jaguar research has been ongoing for years in Sonora's Sierra Madre Occidental and southern Arizona, few hard data exist for the species

Figure 12.1. This adult female jaguar was captured and examined March 30, 2003, on Rancho Los Pavos in Sonora, Mexico. The researchers Emil McCain and Sergio Ávila placed a tracking device on the cat, which managed to slip off its radio collar soon after release. Photo © Emil McCain

in the roughly 140 miles of terrain linking the two areas. Some biologists believe that the intervening mountains constitute a travel corridor that supports transient males, or possibly a small resident breeding population. At least one researcher reported seeing *Panthera onca* tracks in the transit area during 2005, and several area residents insist that they have caught glimpses of jaguars. Following the lead of the Northern Jaguar Project, the Tucson-based Sky Island Alliance has built cooperative relationships with landowners in borderland Sonora in order to encourage conservation research. Ávila, the project's director, supports use of camera traps and hair snags while also looking for feces, tracks, and other physical evidence of jaguars.

Photo-documentation in the United States since 1996 strongly suggests that individual male jaguars are dispersing via insular mountain ranges that straddle the border at random intervals roughly between Sasabe, Arizona, and Antelope Wells, New Mexico. No one knows for certain *why* these cats are on the move. It could be a hunting strategy or a response to overcrowding somewhere else. Some suspect that global warming is sending more of these and other neotropical species (including coatis and peccaries) north, while a few experts believe that the northernmost jaguars are year-round U.S. residents, dispersing east and west above Mexico or simply reoccupying a former homeland by taking advantage of a large prey base and well-managed habitat. Still others feel that such felids are being forced into new territory as a response to hunters in Mexico who routinely shoot them—and the game they depend upon.

"Jaguars are persecuted in Sonora, so males wander in every direction in search of a territory with a female," biologist David E. Brown speculated during our interview at his Arizona State University office in Tempe. "Cattle raising and jaguar conservation are currently incompatible in Sonora."

One locus of hope for *Panthera onca* is a forty-five thousand–acre property between the Río Aros and the Río Yaqui called the Northern Jaguar Reserve, the largest big-cat sanctuary in northern Mexico. It was created in 2003 through purchase of the private Rancho de los Pavos and expanded in 2008 by acquisition of the adjacent, much larger Rancho Zetasora. The Mexican nonprofit Naturalia holds title to the reserve, acquired with donations from private citizens and organizations.

Naturalia's director, Óscar Moctezuma, a scientist and government consultant, estimated in 2007 that about 80 to 120 jaguars had remained in northern Sonora. It was his idea to set up collection cans in Mexican zoos—visited by twenty-five million people a year—for donation of pesos in support of jaguar conservation. The initiative excited thousands of schoolchildren, whose contribution of coins added up to big money over the years. So much positive publicity was generated that many youngsters now regard the jaguar as a symbol of their country's growing conservation movement.

Naturalia is partnered in its Sonora sanctuary with the Northern

Jaguar Project, based in Arizona. Establishment of the reserve was considered a particularly important coup, since Rancho Zetasora's eighty-four-year-old owner was a man who proudly admitted that he had "hated jaguars" and for decades had ordered cats on his property to be shot on sight.

"We hope local people can now make more money off live jaguars than dead ones," the project science coordinator and anthropologist Peter Warshall told me. "Jaguar poaching remains a very serious problem in this part of Mexico." Warshall, citing research by Carlos López González and NJP studies, estimated that twenty-four jaguars were killed between Los Pavos and the Arizona border from 2003 through 2008.

"The greatest hope for saving U.S. jaguars lies in immediately protecting the northernmost remaining population [in Mexico]," declared Craig Miller, the southwest representative of Defenders of Wildlife and vice president of the Northern Jaguar Project. "The seventy-square-mile property will help accomplish just that."

In Sonora, the Northern Jaguar Project and Naturalia are deeply involved in community-based conservation and education programs that benefit local people. Their ongoing "camera survey contest" pays ranchers and cowboys around five hundred dollars for each photograph of a live local jaguar. After a lesser amount was offered for ocelot pictures, eight photographs appeared in short order. Mountain lion and bobcat photos also earn payments. Cowboys and landowners of the area are recruited into a "guardian stewardship" program whereby they become partners in wildlife conservation by attending to camera traps.

However, all is not secure for Sonora's jaguars. Rumors circulate in political and industrial circles that one or more dams may be built in the heart of Sonora's backcountry. (The Aros is the largest undammed river in northern Mexico.) A series of open-pit mines in the region also has been discussed, and a gold mine now operates upstream. Such development would require that roads and structures be built, with relatively undisturbed habitat likely to be damaged or polluted. More people would be drawn to what is now a sparsely populated area accessible only via dirt tracks and horse trails. The potential impact of such projects on area wildlife is unknown.

Despite these ongoing challenges, according to Octavio Rosas-Rosas and other biologists, a growing segment of Mexico's rural population has been persuaded to follow Sonora's lead and climb aboard the jaguar preservation bandwagon. As a direct result, some cattle growers have voluntarily pulled their livestock from remote and marginally profitable areas, or taken mother cows and calves out of prime big-cat country. Inspired by their owners, several large Mexican ranches have, in effect, become private sanctuaries for *Panthera onca*. "The jaguar is an important part of our heritage," one landowner assured me, lips curling into a frown. "It would be an offense to our country to let this animal disappear."

When I asked the man what he thought the best approach to saving the cat might be, he removed his Stetson and ran fingers through matted strands of graying hair.

"We need to work together," the rancher said, replacing his battered hat and breaking into a hopeful smile. "And I know we can have a good time while we're doing it. That's the *mejicano* way."

Emil McCain grew up an only child on a Colorado ranch that bordered Rocky Mountain wilderness. He disappeared into the forest almost daily during childhood and remembers seeing his first mountain lion at age ten.

"Emil is one of those people born 150 years too late," believes T. Luke George, McCain's supervising professor at California's Humboldt State University. "He has skills at tracking that few people in the world have." McCain relied on that expertise to study felids in Central America, Mexico, and the western United States.

"Jaguars are powerful animals and their presence [even when unseen] is overwhelming," McCain told me. "There's a sense of not feeling like you're on top when there's a jaguar around."

Through a series of conversations, articles, and e-mail messages, McCain helped me understand that carnivore conservation strategies are situation specific. A successful approach along the arid U.S.-Mexico desert would probably be inappropriate if applied in the humid forests of southern

Mexico, for instance. Plans tailored for lightly populated Belize might fail in less developed Nicaragua or mountainous Colombia. The only constant among such nations is *Panthera onca*'s status as a veritable "top cat," an animal whose presence in an ecosystem suggests that a wide variety of other plants and animals are also extant. A prevailing scientific argument for jaguar conservation is that such a keystone species helps lock together a geographical ecosystem. Take it away and imbalances ensue; help it come back and those lost balances restore themselves.

A month before our first interview, McCain and his colleagues had made plaster casts of jaguar tracks discovered in a rocky badland an hour's drive south of Tucson. Collection of other tracks in subsequent years, along with camera-trap photos, marked a milestone. According to the wildlife biologist Bill Van Pelt, then of Arizona's Department of Game and Fish, they provided incontrovertible evidence that at least one jaguar—at least part of the time—called the American Southwest home. Between 1848 and 2008, Van Pelt told me, the total number of well-documented individual jaguars seen in Arizona was eighty-four: "This includes adult females and groups of animals, which suggests a [breeding] population in the United States until the early 1960s."

By early 2009 investigators had documented more than eighty jaguar "events" involving four or five individual cats within Arizona and New Mexico during the preceding thirteen years. This included scores of photographs, ten scat samples, nine sets of tracks, three visual sightings, and two video recordings. As a result, the Jaguar Conservation Team recommended more cooperation with Mexican, Native American, and U.S. Border Patrol authorities as part of recovery efforts. It also urged New Mexico to bring its penalties for killing a jaguar in line with Arizona and U.S. guidelines, which make the crime a Class 2 misdemeanor carrying a first-time fine of $2,500, plus up to $72,000 in civil penalties and possible jail time. (At least two private organizations in the region also have posted dollar rewards for information leading to any prosecution for killing a jaguar.)

Figure 12.2. Camera traps placed along this naturally occurring funnel in an Arizona canyon near the Mexico border have taken numerous pictures of jaguars and other animals as well as undocumented immigrants from Latin America. Photo © Emil McCain

Halfway through our two-hour meeting, McCain dropped a bombshell. He and Sergio Ávila had been cautious until then in answering questions posed by the man writing the book. Perhaps my zeal prompted the researchers to relax.

"I don't think jaguars ever left Arizona," McCain declared, swallowing a last bite of turkey sandwich. "I believe they've always been here. We just stopped seeing them."

McCain's view is contrary to the opinions of Alan Rabinowitz of

Panthera and David E. Brown of ASU, among others. "The jaguars we've seen in the United States since 1996 are single males wandering up from Mexico," Brown had insisted during our conversation. "In the greater scheme of things, they're insignificant until females are present. If one is serious about establishing a jaguar population in the United States, one would want to reintroduce females."

Despite cowriting, with Carlos López González, a book about border-area cats, Brown has a dismissive take on the subject. "The jaguar is a fantasy animal," he told me. "It's like the Abominable Snowman. . . . People see it because they want to and I resent spending so much public money on chasing down bogus sightings." Brown dismissed attempts at *Panthera onca* management by the Jaguar Conservation Team and other groups as "politics and nothing more."

McCain vehemently disagrees with Brown, whose expertise as a biologist the younger man is quick to praise. He speculates that more jaguars were seen in Arizona and New Mexico before 1950 than since because people once traveled more often into remote areas. "Cowboys used to live and tend stock out on the range," he maintained. "Prospectors wandered around looking for minerals. Homesteaders broke ground. In the Southwest, these things pretty much stopped after World War II."

In 1999 the Arizona Game and Fish Department gave assurances that it would not advocate, support, or allow jaguar reintroduction. A report prepared in 2006 by the wildlife agencies of Arizona and New Mexico did, however, recommend conservation of key jaguar habitat, particularly mountain ranges connecting these states with Mexico. Such proposals have met strong resistance from some public-land ranchers, who fear reintroduced jaguars may develop a taste for local cattle and thus threaten their livelihoods. Bucking the trend is the iconoclast Will Holder of Eagle Creek, Arizona, who tolerates killing of his calves by wild animals and recoups his losses by marketing a premium line of "predator-friendly" beef. It has been suggested by some ranchers that such beef might also be attractive to consumers willing to pay extra in order to compensate for any calf depredation.

As of early 2009 only one instance of a jaguar livestock kill had been confirmed in the Southwest since the Glenn sighting of 1996. The Arizona rancher was paid five hundred dollars for his 2008 loss.

≈≈

Perhaps the biggest short-term threat to U.S. jaguars has nothing to do with the fears of ranchers or homing instincts of felines but rather with an unnatural set of border barriers and monitoring techniques. In the years following the terrorist attacks of September 11, 2001, illegal immigration from Mexico became a hot-button political issue. As a result, surveillance by the U.S. Border Patrol and other agencies has intensified along the nation's southern boundary, posing a potential threat to many wild creatures that routinely cross the frontier.

"It will be a shame if [the Border Patrol] builds a tall fence along the whole border," ASU's Brown told me, referring to the seven hundred–mile physical barrier authorized by Congress in 2006. "The jaguars aren't the only animals that won't get across. It will affect mountain lions, deer, bear, opossums, and other mammals." Nature-related tourism and hunting in the region, he said, also might be affected adversely.

The proposed fence, estimated to cost up to $1.2 billion, could seal much of the U.S.-Mexico border with an impenetrable barrier. If the fence extends through *Panthera onca* habitat, Peter Warshall of the Northern Jaguar Project predicted, "you could kiss the [Southwest] jaguar goodbye."

Enabling legislation allowed the Department of Homeland Security to redirect funding from the wall to less obstructive measures, including a high-tech "virtual fence" that would use digital electronics to monitor activity. This approach would be less likely to run afoul of the North American Commission for Environmental Cooperation, set up to protect ecosystems along U.S. borders with Mexico and Canada. But instead, construction of physical barriers began in jaguar country in mid-2007, causing major problems for wildlife researchers.

The nonprofit Center for Biological Diversity was one of several environmental organizations that criticized the Border Patrol and the Department of Homeland Security for carrying out activities that "fragment jaguar habitat and likely restrict jaguar movements." Besides tall fences, the center's complaint list included stadium-type lighting, destruction of vegetation, airplane surveillance, and construction of heavily patrolled roads. Legal challenges by this group and others, including a coalition of regional lawmakers, were thwarted by federal agencies, including the Fish and Wildlife Service, with the response that protection of jaguars in the United States was not essential to the felid's survival as a species. In June 2008 the U.S. Supreme Court refused to hear a case challenging the federal government's contention that it can waive environmental laws in the interest of national security.

❧❧

The Borderlands Jaguar Detection Project has installed more than fifty cameras near the Arizona-Mexico border. From 2004 through early 2009 the project photographed a single jaguar, the cat nicknamed Macho B. "We've taken fifty-two pictures of him," Childs told me in a three-way discussion with his wife, Anna Mary, at their five-acre spread near Amado, Arizona.

Jack Childs is sure that Macho B has made trips into Mexico—"We've photographed him at the fence"—but the researcher wants to learn whether this mature specimen travels as far as the Sierra Madre, home of the nearest known females. A straight-line route would be nearly two hundred miles, a travel distance not unheard of for highly motivated male jaguars.

The answer to such questions may yet be known. On February 19, 2009, surprised scientists studying bears and mountain lions near the U.S.-Mexico border southwest of Tucson unexpectedly snared Macho B, the much-photographed Arizona jaguar videotaped by the Childs party in August 1996. After sedation, the cat was fitted with a satellite-monitored radio collar that reports its exact location every three hours. Macho B, now at least fifteen years old, weighed 118 pounds and was pronounced "healthy and hardy." Unfortunately, the jaguar suffered kidney failure and stopped eating soon afterward. Macho B was recaptured and euthanized on March 2, 2009.

"You cannot generalize from [one cat] to what all jaguars do," cau-

Figure 12.3. From left, Jack Childs, Anna Mary Childs, and Emil McCain of the Borderlands Jaguar Detection Project examine jaguar tracks in southern Arizona. The three have been in the forefront of U.S. jaguar conservation and research for many years. Photo © Mitch Tobin

tioned Ávila. But McCain countered that "if you're going to provide a sound future for the jaguar [in the Southwest] you've got to know its needs."

Camera-trap photographs of jaguars in the American Southwest dwindled after mid-2007, as construction of border barriers and physical surveillance intensified along the international boundary. Macho B was glimpsed once in mid-2008 and again in early 2009, but no other jaguar sightings were confirmed. Researchers seemed optimistic, however, that the first radio-collared jaguar in U.S. history might lead them to other resident members of the species.

CHAPTER THIRTEEN

"To Ensure Our Namesake Is Protected"

MY LATE COUSIN THE NEW MEXICO oilman Bob Enfield dearly loved fast, high-performance cars. I rode shotgun with him as a kid and watching with astonishment as his eight-cylinder sedan's speedometer crept above one hundred miles per hour. I had never been driven so fast and was sure that somewhere between Vaughn and Roswell we would lift off the asphalt and soar across the Llano Estacado like a flying saucer.

When I was fourteen, Bob became one of the first in his state to buy the spiffy 1965 Ford Mustang, an instant classic. I traveled in it halfway across Texas, drawing whistles of appreciation for five hundred miles. Thirty years later, my cousin impressed the family by offering spins around Santa Fe in a gleaming green V-12 Jaguar coupe, obtained by special order through a dealer in Phoenix.

"Wow," exclaimed my father, Don, a buyer of no-nonsense station wagons and practical commuter cars, "will you look at that!" While stationed in England with the U.S. Air Force after World War II, Don had taken some British Jags for joyrides and thereafter jokingly referred to any powerful, good-looking car as "a Jaguar." His nephew-in-law's Jag purred like a contented housecat. Don needled Bob about the speeding tickets his stylish ride produced, but my cousin took this teasing in stride. The lunging chrome cat on the long, shapely hood, he figured, was not designed to stand still—and traffic citations were a sort of excise tax.

"I love my Jag," Bob said, beaming behind the wheel of a spit-polished machine. "This car is so much fun to drive—and it's damn gorgeous." I never heard anyone contest these claims. But I also never knew how a charismatic cat became a cultural icon, synonymous with a classy ride. I contacted the Jaguar Motor Company, which put me in touch with its official historian.

⌐▄▄⌐

The Jaguar's gestation as an automobile began in 1922 in a drab factory operated by a British outfit called the Swallow Sidecar and Coachbuilding Company. A manufacturer of motorcycle sidecars, Swallow owner William Lyons, who cofounded the firm on his twenty-first birthday, was described by the *London Telegraph* as "fiercely ambitious." The scion of a wealthy family decided to expand into the auto business during the car-crazy decade preceding World War II. After buying out his partner in 1934, Lyons switched the company name to SS Cars Limited and produced several two-seat models at Coventry in the "S.S." series, believed to be an abbreviation for Standard Swallow.

In 1935 Lyons settled on Jaguar as the name for his newest marque, reportedly choosing this moniker from among several choices presented by Nelson Advertising, then under contract to provide Swallow's publicity and marketing services. Lyons is said to have liked Jaguar as a name because of its association with the understated strength, elegance, and smooth agility of the charismatic jungle cat. A 1936 edition "SS Jaguar 100" car was the first to display the appellation and updated styling. The following year the automobile added its now-famous "leaping jaguar" hood ornament, designed by the artist and sculptor F. Gordon Crosby.

London joined the fight against Nazi Germany in 1939, and Lyons manufacturing was diverted to the British war effort from 1940 through 1945, after which the Jaguar model reappeared. In February 1945 Swallow's name was changed formally from SS Cars Ltd. to Jaguar Cars Ltd. (A company spokeswoman told me that there is no truth to the rumor that the earlier title was abandoned because of negative associations with the Nazi regime's Schutzstaffel, the secret police commonly referred to as the SS.)

For Jaguar's postwar iteration, Lyons sought to create ever more refined, high-performance automobiles, giving his new models a decidedly feline look and the capacity to accelerate within seconds—like a pouncing cat

Figure 13.1. Sir William Lyons, the British founder of
Jaguar Cars Ltd., poses next to one of his company's
stylish XJ models. Lyons, who died in 1985, picked the
jaguar name because of its association with elegance
and power. He sold Jaguar Cars to an English corporation in 1966. The company was sold by Ford to India-
based Tata Motors in 2008. Photo courtesy of Jaguar
North America

—all the way to 150 miles per hour. Some maintain that these late 1940s
editions were the first true world-class sports cars, ushering in a grand era of
sophisticated British motoring that included swift, fun-to-drive sedans from
Aston Martin, Austin-Healey, Cooper, and other manufacturers. Perhaps
in recognition of the entrepreneur's success in getting high-performance
British sports cars noticed by affluent consumers worldwide, Queen Elizabeth II knighted William Lyons in 1956.

Through the 1950s and 1960s the Jaguar was respected (and envied) for its athletic panache and phenomenal engineering. Between 1951 and 1957, Jaguars won five times at France's prestigious Le Mans Grand Prix. The Jaguar name became associated with an elite mix of race drivers, movie stars, political leaders, and aristocrats. Naturally, the higher the price tag the bigger the bragging rights. Nothing gave a big-name personality more snob appeal than pulling up to a public event in a fancy Jaguar. Conspicuous owners during this era included manly matinee idols Clark Gable (whose 1954 XK-120 was the first exported to North America) and Steve McQueen (who drove a widely coveted 1956 XK-SS). Other celebrity drivers were Frank Sinatra, Peter Sellers, Nelson Rockefeller, John Wayne, Elvis Presley, Lauren Bacall, Humphrey Bogart, King Michael I of Romania, and the Shah of Iran. Some owners made unusual modifications. For example, mystery novelist Mickey Spillane, creator of the best-selling Mike Hammer detective series, installed in his Jaguar the safety belt from a Mustang P-51 military plane Spillane had used to train fellow pilots during World War II.

In 1966, Lyons merged his company with the larger British Motor Corporation in order to improve Jaguar's manufacturing facilities and strengthen its financial resources. But he and other stockholders were distressed to see Jaguar share prices tumble following the merger. BMC was taken over by British Leyland Motor Corporation only two years later. Lyons stayed on as chief executive officer until his retirement in 1972. He continued as a consultant on all new Jaguar models until his death in 1985 at age eighty-four.

Jaguar passed out of British control in 1989, when Detroit-based Ford Motor Company bought the automaker for a reported $2.5 billion. Autos continued to be designed and made in England, with a stated corporate vision of producing "beautiful fast cars that are desired the world over." Ford promoted Jaguar as a vehicle of uncommon strength and craftsmanship, producing one model—the XJ-2220—that sold for $500,000 and achieved a top speed of 212 miles per hour. More generic editions retailed for around $50,000 in 2008, placing them well within the luxury class.

From an economic point of view, Ford's investment in Jaguar was an acknowledged disaster. According to the *Detroit News,* company officials conceded the unit lost money no matter what strategy was employed; in 2007 losses were put at $715 million. The following year, faced with declining sales in other divisions, Ford quietly put Jaguar up for sale. The brand was purchased in March 2008 by Tata, India's biggest car company. Reported terms were $2.3 billion plus various union commitments and financial guarantees.

A chrome-plated ornament on the hood of a Jaguar continues to symbolize one of the industry's most sought-after cars. A recognizable branding icon worldwide, the so-called Leaper adorns most models and has changed only slightly since its introduction in 1937. The embellished feline is intended to stress smooth lines and athletic grace, its open jaws and retracted ears to evoke fearless confidence.

The first major change in the Leaper came in 1955, when its rear legs went from crouched to outstretched. Later versions brought a more stylized head and an upward tip to the tail, which otherwise lies flat against the jaguar's rear legs. In recent years laws enacted to protect pedestrians from injury have shifted placement and attachment of the Leaper. On some models and on Jaguars sold in the Netherlands and a few other countries the ornament has disappeared altogether. In the United States it is designed to swivel or to snap off easily in the event of an accident, thanks to a spring-loaded mechanism placed beneath the hood.

The Leaper sells as a separate accessory for $250 or more, depending on its pedigree. It is a popular collectible among Jaguar aficionados, and highly desirable versions sell for thousands of dollars. A less prestigious add-on is the snarling jaguar face known as the "Growler," a two-dimensional symbol mounted above the distinctive front grille and on side panels of some models.

For millions of affluent, status-conscious consumers around the planet, a Jaguar automobile continues to represent impeccable standards of wealth, taste, performance, and beauty. In fulfilling that last category for

Figure 13.2. Variants of this chrome-plated ornament, dubbed the Leaper, have embellished the Jaguar's hood since soon after the automobile first rolled out of an English factory in the 1930s. Photo courtesy of Jaguar North America

a 2007 concept model, the company created a particularly shapely edition based on the female form of British actress Kate Winslet. The star of *Titanic* and other films was judged by the car's architect to possess the shape of the "ideal woman." Scottish engineer Ian Callum, according to the *Times of London,* "designed the new XK body with [Winslet] in mind." When asked how she felt about having a car inspired by her voluptuous figure, Winslet told U.S. television talk-show host Jay Leno: "It's kind of flattering

. . . [but] the headlights aren't big enough," and it could use "a bar under the dashboard."

❧

There's more to a Jaguar car than flash, however. For decades, a portion of company earnings helped its namesake survive in the wild. Beginning in the mid-1980s, the manufacturer donated millions of dollars for scientific research and environmental education related to the flesh-and-blood animal. The car company began its support of jaguar conservation through an initiative launched by Jaguar Canada, whose executives showed particular concern about the disappearance of *Panthera onca*. A substantial financial contribution by Jaguar Canada helped the World Wildlife Fund, in cooperation with Alan Rabinowitz and the Belize Audubon Society, to create the world's first jaguar sanctuary in the Cockscomb Basin of southern Belize.

The company went on to donate one million dollars over a five-year period to the New York–based Wildlife Conservation Society for expansion of the nonprofit organization's jaguar studies in Guatemala, Mexico, Costa Rica, Belize, and other countries. The automaker also helped underwrite an unprecedented 1999 Mexico meeting of prominent jaguar researchers that yielded panel discussions, a bilingual reference book (*Jaguars and the New Millennium*), and a commitment to help maintain safe corridors between major jaguar habitats from north to south. "Without this type of [financial] support," attendee Rabinowitz noted approvingly, "we're not going to be able to save these animals."

In April 2003, Jaguar North America created the Jaguar Conservation Trust as a way to support a wide range of projects devoted to saving wild jaguars in several countries, including the United States. In the words of then–executive vice president of sales and marketing Richard Beattie, it was time "to take our conservation efforts to the next level and ensure that the future of our namesake is protected." The trust's first conservation consultant was actress and environmental activist Stefanie Powers, who described her involvement as "extremely satisfying."

Despite heavy corporate losses, the car company continued to fund new programs between 2004 and 2008 through the trust, making small grants to individuals and organizations. In addition to outright donations, Jaguar North America forwarded a percentage of proceeds to the trust from three popular items in its Jaguar Collection merchandise catalogue. (The unit sells clothing, gifts, toys, and collectibles that bear the Leaper logo.)

Not all funds went to field research. Jaguar Cars donated three million dollars to Britain's Chester Zoo for state-of-the-art educational exhibits on rain forest and jaguar preservation. The northwest England facility annually attracts more than one million visitors, including some 70,000 students. Meanwhile, in San Jose, California, Jaguar North America and a local dealer underwrote a celebratory "coming out" party for a 17-month-old jaguar cub named Sophia at the Happy Hollow Zoo. The sponsors paid for a new exhibition space and other amenities for the young cat, which succeeded Jezebel, a popular jaguar who died at 23 of age-related illness.

It was unclear in early 2009 whether support for jaguar conservation would continue under the company's Indian owners.

While the Jaguar automobile is a modern technological phenomenon, the feline "perfection" it represents has long been embedded in the human imagination. "Ever since our kind could pick up a paintbrush or a carving tool," wrote Diana Landau in *Clan of the Wild Cats*, "the likenesses of cats have been rendered. To call the [human] preoccupation with big cats 'fascination' would be a tremendous understatement."

The obsession with cats among Western civilizations has prevailed for millennia. The ancient Egyptians are sometimes credited with domesticating today's house cat (*Felis catus domesticus*) from a wild desert felid (*Felis silvestris lybica*) some ten thousand years ago. Other sources speculate that such cats basically domesticated themselves by choosing to live near people—and the rodents they attracted—in the earliest days of grain farming. Whatever the case, we know that during the pyramid-building era along the Nile, cats came to be worshiped as presumed offspring of the sun god, Ra. A goddess who took the form of a cat, Batus, was believed to rule the Egyptian home.

In contrast, most of the human veneration of jaguars has been at a distance. Although the effort is rarely successful, an occasional daring soul tries to keep a jaguar in captivity as a high-status companion. For example, the nineteenth-century English poet Lord Alfred Douglas had an Aunt Florrie—a well-known eccentric—who kept a pet jaguar that reportedly ran amok and killed several of Queen Victoria's favorite deer. Across the Atlantic, zoos of Latin America long have welcomed jaguars brought to them for donation by chagrined owners—often, it seems, minor public officials or rural politicians—who obtain them as cubs only to find the cats dangerously unmanageable in adulthood. The turnover continues to this day, as a search of the Internet will reveal.

Would-be owners might have done well to look up "jaguar" in the *New International Encyclopedia,* 1906 edition. "These have sometimes been tamed," according to the entry, "but this species is perhaps the most savage and intractable of the great cats, and the kittens become dangerous with increasing years." Setting hyperbole aside, the underlying message rings true: unusual is the captive jaguar that can truly be tamed.

The actor Marlon Brando is said to have succeeded. While living in the Santa Monica Mountains above Hollywood, Brando reportedly kept a pet black jaguar in his home. This unverified story claims that a lover of Brando's once balked at the actor's suggestion that the couple share his bed with the full-grown cat. It allegedly curled up regularly at Brando's feet on what must have been a very large mattress.

The movie star was perhaps competing with flamboyant Spanish painter Salvador Dalí, who traveled widely with a muzzled, semitame ocelot he kept on a leash. The artist—hardly publicity shy—sailed on cruise ships and rode in limousines with the spotted cat, with which Dalí was often photographed. "If his ocelot urinated on one of Dalí's paintings," a knowledgeable second-hand source told me, "the artist always made sure to charge the buyer more money for that particular piece of cat-sprayed art."

Public fascination with exotic big cats as pets is borne out in a number

of Hollywood films featuring such felines. In 1938's *Bringing Up Baby,*
for instance, Katharine Hepburn co-stars with Cary Grant as the wealthy
keeper of Baby, an adult black leopard—hardly a "baby"—whose escape
leads to screwball comedy and the inevitable union of the mismatched lead
characters. A good-natured and well-trained captive leopard named Nissa
played Baby in the film.

Through the centuries a wide range of owners have kept nondomestic
cats as pets, though experts contend that the animals' wild instincts and
behaviors cannot be fully subdued. What's more, the bigger felids seldom
form empathetic or compassionate bonds with humans. The relationship is
not considered an emotional one, but rather one of convenience. Whereas
an owner may feel love for his or her exotic feline, experts say the animal
is more focused on the food and security the owner provides. "You cannot
tame a jaguar," I was assured by a professional trainer with years of experience in raising large captive cats: "An adult jaguar can easily overpower you
if it so chooses." Despite such problematic realities, the desire for private
ownership of large felids seems ongoing and universal.

In North America, the most widespread borrowing of the jaguar's charismatic image and admirable attributes is by athletic teams. The formidable
cat is a perennially popular team mascot. Hundreds of football, basketball,
soccer, track, hockey, and baseball organizations throughout the United
States, Mexico, and Canada incorporate jaguars in their names, logos, advertising, merchandise, and various kinds of imagery.

The Jacksonville Jaguars, for example, became the National Football League's thirtieth franchise in 1993. Fans chose the name by voting
in a team-sponsored competition that offered three other options: sharks,
stingrays, and panthers. Jaguars were picked despite the empirical reality that the species probably has not occurred in Florida for thousands of
years. (Fossils confirm a prehistoric presence there, perhaps even since
the last ice age ended about 11,000 years ago.) When the Jacksonville team
played its first game, in 1994, it bore as its logo a devilish caricature of a

snarling jaguar with a prominent blue tongue. (A living jaguar's tongue is pink.) The costumed mascot that entertains cheering fans at Jaguar games is known as Jaxson deVille.

Franchises such as the Jacksonville Jaguars generate huge revenues through the sale of branded merchandise—from hats and socks to mugs and toys—that trade on their namesake's image. My research unearthed no evidence of a sports team directing any funds toward preservation or study of real-live jaguars. However, at least a couple of teams have kept four-footed versions of their symbol in cages, to be displayed to students, faculty, and alumni for promotional purposes.

As an example, for thirty years Southern University housed and fed a pet jaguar as a live mascot representing the Baton Rouge school's athletic program, despite criticism that the cat's 418-square-foot enclosure—about the size of a one-car garage—was too small for such a large animal. The tradition ended in 2006 with the death of Lacumba II, the last in a series of SU jaguars. During the late 1960s, the University of South Alabama also kept a jaguar. The live mascot was inexplicably named Mischka, a Russian term for "bear." SAU rid itself of the felid after someone left its pen open and the cat roamed freely around the school. It was captured without incident. In 2007 a family member forwarded to me two photographs of a captive jaguar taken at a Mississippi zoo. The cat reportedly was once owned by SAU. Indiana University–Purdue University at Indianapolis, which also claims the jaguar as its good luck charm, eschews a live animal in favor of two costumed versions, Jinx and Jawz. Spelman College in Atlanta and the University of North Texas at Dallas prefer to bring out cartoonlike, costumed jaguar figures at their games.

The adoption trend extends south of the border as well. A popular Mexican soccer league team is Jaguares de Chiapas, a reference to the once mighty jaguar stronghold in the southernmost rain forests of Mexico. Several prominent rugby teams in South America also include jaguars in their names or emblems.

≈≈

In the early twenty-first century, the enthusiastic use of the jaguar as motif, icon, logo, brand, and charmed name continues unchecked. Over the years I have kept an informal record of some of the myriad ways commercial, governmental, and educational interests have appropriated the name *jaguar*. By my count, it has been all or part of the name for at least one of the following: nightclub, restaurant, jet fighter plane, computer operating system, film distributor, movie title, men's cologne, marketing service, mining company, gas station chain, cartoon character, rock music group, dry cleaner, advertising agency, video game console, transport company, investment firm, used car dealership, courier service, bus line, electric guitar, coat of arms element, and any number of products. Entrepreneurs seldom connect the latter with jaguar conservation, but an exception was reported in 2007 when a $30 Jaguar Vodka was unveiled that used sugar and water from Barbados. "Some of the proceeds," promised promoters, were sent to preserve jaguar habitat through a nonprofit organization in Brazil.

The term *black panther*—a common misnomer for the uncommon melanistic jaguar or leopard with uniformly dark fur—has been applied to several political organizations and music groups, including the first band in which contemporary rock musician Ozzy Osbourne played. However, the most famous borrowing came in northern California during October 1966 when activists Huey P. Newton and Bobby Seale formed a political organization called Black Panthers for Self Defense. The name, later shortened to Black Panthers, apparently was chosen to reflect the symbolic power embodied in the cat and its perceived pride in being black.

The Black Panther organization was founded in Oakland at a time when that majority-minority city reportedly employed only 16 African-American law enforcement officers, out of 661 total. Allegations of police abuse in Oakland were common. The group espoused the right of African-Americans to carry arms as well as to defend themselves against police brutality and other forms of institutional oppression. The Black Panthers allied themselves with poor and working-class residents through promotion of free food, clothing, recreation, and health care programs as well as affirmative employment, civil rights, and education initiatives. The group's appeal remained limited in part because some influential members were militaristic in their tactics, guided by a confrontational ideology described openly as "Marxist-Leninist socialism" and "proletarian internationalism."

The Black Panthers made headlines at the 1968 Summer Olympics when two medal-winning African-American athletes raised black-gloved fists above their heads in a Panther-associated salute to "black power." The Mexico City incident, which resulted in immediate censure of the track stars, marked the beginning of the end for the group as an effective force for socio-political change. Instead it became a lightning rod for authorities eager to crack down on radical activists. During the late 1960s an estimated thirty-nine Black Panthers were shot by police and three hundred members put on trial. In 1970, Federal Bureau of Investigation Director J. Edgar Hoover labeled the organization "the greatest threat to the internal security of the United States." A combination of FBI infiltration, high legal defense costs, and leadership problems led to disintegration of the Black Panthers during the early 1970s.

In Latin America during the late twentieth and early twenty-first centuries, jaguars drew renewed interest among artists, playwrights, commercial enterprises, and community leaders. One example of this revival enjoys an honored place on my desk, near the computer I use to research and to write about *Panthera onca*. The simple wooden carving is about nine inches high and three inches wide. It is an imaginative rendering of a jaguar seated upright on its hind legs and extending a forepaw in the "let's shake hands" gesture of a trained dog. This particular jaguar has reddish flanks and a yellow-orange belly. Its head, back, and legs are dotted with black spots. A Maya artisan in Mexico's state of Yucatán carved and painted the decorative souvenir. My travel-writer friend Nicky Leach purchased the sculpture in a *mercado* and brought it to me as a gift. I am comforted in knowing that at this moment throughout Latin America, thousands of equally fanciful jaguars are either perched on office desks or sitting in shops, businesses, museums, schools, and homes.

Mexico's fascination with jaguar imagery is pervasive and enduring. The love of Mexicans for *el tigre* can be partially explained by the customs and mythology of the Aztecs, the highly developed civilization of Mexico's central plateau that immediately preceded—and violently resisted—Spanish colonization. Though long gone in its pure form, the Aztec culture is one with which millions of *mestizo* Mexicans still closely identify. In southern Mexico, those of Olmec, Zapotec, or Maya ancestry show similar pride and attachment to their own distinct heritages. But it is the Aztecs, at their shimmering height when cruelly conquered, who hold a special place in Mexico's heart—and sharing every beat of this *corazón* is the jaguar.

CHAPTER FOURTEEN

"Siga el Pisto"

MY QUEST NEXT TOOK ME TO the cool highlands of Guatemala, a hodgepodge of coffee plantations and pine forests, where I loosened my rusty Spanish at an Antigua language school. Within a few days of marathon instruction I was sufficiently fluent to discuss jaguars with my patient teachers, Maritza and Abelino. They were surprised to learn that these cats still existed in their country. Maritza, a bookish *Antigueña* with an interest in feminism and politics, reacted with wide-eyed astonishment: "I thought the last of them must have been slaughtered by ranchers and poachers long ago." Abelino, a quietly earnest fellow who commuted from neighboring Pastores, adopted a wistful look: "As a boy, I heard about *jaguares* appearing like phantoms on the slopes of the three volcanoes that tower over us. I also remember seeing tracks of deer along the river. Now," sighed Abelino, "all wildlife is gone from here except for some common birds and a few nervous rabbits."

We agreed that this turn of events was *"que lástima"*—too bad—but not surprising given the unrestricted hunting throughout Guatemala. The *señora* in whose home I was boarding was more sanguine, shrugging her shoulders and pointing in the direction of two parrots her deceased husband had captured in the forest. The wing-clipped birds spent their days clinging to a wooden roost. "When my *esposo* was alive," Sylvia said matter-of-factly, "he was stationed for weeks in the jungle as an energy-company executive and went after everything that moved. So did everybody else."

After two weeks of conjugating verbs and parsing pronouns, my host and instructors wished me *buena suerte* and sent me off to Guatemala City. From there I made a bone-jarring eleven-hour bus ride deep into the Petén, the steamy lowland of the subtropic north. I sat next to a young tourist from Poland who spoke no more than a dozen words of English and even fewer of Spanish. It amazed me that a single, fair-haired woman would undertake such an adventure in a country with a shockingly high crime rate. I decided that in this case ignorance might be bliss, since my seatmate couldn't read my newspaper's account of an average of twenty murders a week ravaging this West Virginia–sized country. But when I showed Petra a photo of a jaguar, her face lit up. "Tiger!" she exclaimed.

Our weary bus overflowed with displaced Guatemalans and immigrating Salvadorans starting new lives on the jungle frontier. Once a hotspot of biodiversity, the rolling subtropical plateau was slipping into ecological blandness. I saw results of peasant resettlement outside my window: a mélange of irregular vegetable patches and cattle pastures, punctuated by simple cement-block houses beyond which clusters of children played stick games among pigs, dogs, and chickens. In the fourteen years since my first trip through the Petén, tens of thousands of acres of forest had been cleared.

More troubling was the dark mood I stepped into upon arrival in Santa Elena, where a chorus of voices from an outdoor café warned me not to walk the streets on this raucous Saturday night. "You will be robbed," an old man playing dominos assured me. "Or worse," interjected his partner. Gangs of petty criminals and drug dealers, it seemed, ruled the night.

I hailed a taxi for the short ride to Flores, the tiny (and comparatively safe) Petén capital ensconced on a perfectly circular island in Lake Petén Itzá. Connected to the mainland by a narrow causeway, Flores long has been popular for the lodging it provides tourists making day-trips to the magnificent Maya ruin of Tikal, about an hour away. Flores itself is the modern manifestation of Tayasal, the last remaining intact Maya city when conquered with great violence by the Spanish in 1697.

After a full day in Flores, during which I noted the presence of endangered wild game on restaurant menus and an ocelot hide tacked to a wall, I met with Roan Balas McNab, the U.S.-born anthropologist who was beginning

his second decade overseeing Guatemala's office of the Wildlife Conservation Society. The director's desk was shoehorned into a cramped bungalow
wedged along one of the pedestrian-only alleys climbing the steep hill that
dominates Flores.

The slender, blue-eyed, blue-jeaned American was breezily friendly,
but intense, clearly a man who hated to waste time. McNab's first comment
to me after *buenos días* was: "There's a lot of work to be done to change this
world and it's piling higher every minute." He chuckled good-naturedly and
estimated "more than nine hundred e-mails" awaited his reply, accrued
during the seventeen days McNab had just spent "walking in the woods"
near the Mexico border. After exchanging preliminaries and dashing to
his home in order to retrieve field reports and a reference book, we agreed
to meet for lunch the following afternoon. I spent the evening poring over
Spanish-language research papers in which McNab described Guatemalan
wildlife. They suggested that tenuous conditions prevailed for large carnivores. One report after another confirmed that fauna was disappearing
faster than it could be protected.

"Much to our dismay," McNab wrote in a 1999 assessment, "jaguars
have never been studied in Guatemala despite the fact that the country's 1.5
million hectare Maya Biosphere Reserve functions as the core protected area
for possibly the largest remaining contiguous population of jaguars north of
Colombia." Subsequent research by WCS and other groups suggested that
jaguars, which once roamed almost everywhere in Guatemala, were reliably
found in less than one-third of the country, mainly its northeast corner. Another study concluded that jaguar skins were being sold openly in public markets, for prices ranging from $40 to $120 each. How many jaguars remained
in the country was unknown. An educated guess put the number at fewer
than four hundred. Estimates for all of Mesoamerica had ranged between
one thousand and five thousand individuals in the mid-1990s.

In the morning I paid a local boatman to shuttle me across Lake Petén
Itzá to a regional zoo where living examples of Petén jungle fauna were on

Figure 14.1. Seasonal flooding of pathways used by jaguars in tropical forests makes tracking and photography challenging for researchers. The wildlife biologist Carolyn Miller has studied big cats in La Selva Maya since 1993, personally encountering more than thirty jaguars and capturing dozens of others on film. Photo by Carolyn M. Miller

display. As he steered José volunteered that he indulged in eating *faisán, jacalí,* and *venado*—curassow, peccary, and deer—at every opportunity. "Qué sabroso," he declared, smacking his lips. "So delicious!" I pointed out that each of these was also a preferred prey of jaguars, which were fast disappearing from the surrounding forest. José shrugged his shoulders and leaned forward: "I will give you the name of a café in Flores where you can sample all three, plus armadillo, tepezcuintle, and wild turkey."

Moments later we landed on Petencito. My guide showed me first around the tiny island, showing off a few bored-looking animals: a crocodile, several peccaries, and scarlet macaws with clipped wings. A fenced compound on the mainland, accessible by pedestrian bridge, clung to a steep hillside. It held two jaguars, a male and a female. These animals were gorgeous and seemed reasonably well cared for, their bodies trim and coats shiny. The male growled and roared, then gave what appeared to be an affectionate tongue-lashing to the face of the female. I wondered whether they might be siblings. The male sauntered to the shade of a tall tree and sat on his haunches, then picked up his front legs one at a time and licked them clean in the familiar method of a housecat. The female set off on a patrol of the cage's perimeter, a distance of about two hundred yards. She huffed and chuffed, flicked flies away from her ears and eyes, and focused on a trail beaten into the soil. The power of this cat, though restrained by circumstance, was evident in her rippling muscles and effortless motion. I felt sorry, as one often does, at the spectacle of a caged predator, but viscerally excited by simply watching the two cats moving about beneath a luxuriant jungle canopy. Adjacent to the jaguar pen was an empty enclosure. José told me that it once had contained a mountain lion.

"It died," he said, without elaboration.

"Have you ever seen a jaguar in the wild?" I asked.

"Once," he replied. "In Tikal National Park."

"Do *Peteñeros* still hunt them?"

"Yes, they do; for their skin or for a trophy head to mount on the wall. The cubs are captured alive and sold as pets. But only a few are left in the forest now."

We were interrupted by the female jaguar's throaty growl, low and deep. It ended in a rasping cough. The sound shot a chill down my spine.

"What do they get fed here?" I wondered.

"Raw chickens and eggs," came the reply.

A caretaker named Antonio appeared and escorted me up a steep trail to a narrow point between two boulders. He showed me a laminated color photo with a caption that detailed the shooting of a jaguar at this exact location in 1991. "Blood poured from the cat's mouth and left a stain on the ground for many months," Antonio said, "but the mark is gone now. I guess the rain finally washed it away."

My interview subject arrived at Restaurante Guacamaya dressed in khaki pants, a short-sleeved shirt, and sandals. Restless with energy, the flaxen-maned researcher said that he had been burned by journalists and insisted that his more provocative remarks be kept off the record. I agreed. Now middle-aged, the man had spent much of his career in Central America and offered an insider's perspective steeped in on-the-ground experience.

"Maybe the most important thing to know [about Guatemalan jaguars]," McNab began, "is that twelve million people are citizens of Guatemala and that population is probably underestimated by another two million. The nation's annual growth rate is 2.5 percent." With this many people squeezed into a space half-covered by forest, habitat loss was inevitable. The average monthly income, he added, was "less than one hundred dollars, and most Guatemalans have few skills and little education. It's a cinch to sneak in and out of the country, the cost of living is low, and it's pretty easy to find menial work or do subsistence farming." The nation he was describing was one in flux, vulnerable to winds of change its institutions could not control, much less steer.

This added up, in McNab's view, to strong forces working against *Panthera onca*. "We know that jaguars are here," he declared, "but they've become extremely cautious. In ten years of looking I still have not seen one. Oh, I've found tracks, scratches, scrapes, sprays, and feces. One even ran in front of a group of hikers I was with while I was looking the other way. They're very smart animals to live among humans for so long without being seen; I hope that keeps working to their benefit."

Over the next hour McNab described various projects undertaken to study and conserve Guatemalan jaguars. Besides educating local residents about the importance of preserving the cats, much of his organization's work involved camera trapping or patrolling the forest in search of jaguar signs. A strategic goal was to identify blocks of forest containing at least fifty jaguars, since smaller populations may not be viable without protected corridors linking them to other jaguars.

McNab was particularly proud of an innovative program under way in Uaxactún, a village near Tikal. After years of environmental education, residents had been conscripted to report evidence of jaguars and encouraged not to shoot the animals. "These cats have been known to go right inside villagers' kitchens looking for food," said McNab, appearing incredulous. He agreed that in such cases it might be prudent to trap or kill the offending jaguar. "But firing a gun at a big cat may injure it, which can make its behavior much worse." (Two weeks later, I examined the skull of a jaguar that had been shot twice. Pellets from the first shooting, which the cat survived, were still embedded in its forehead.) Roan had spent ten years trying to convince skeptical Peteñeros that hunting should be managed carefully in order to keep game animals sustainable. Modest headway had been made, and more local residents seemed to understand the need for limits.

In one study, Uaxactún hunters were encouraged to bring to the WCS team any feces from large cats they came upon in the forest. After processing, the material was carefully examined in the United States in order to determine what prey animals the jaguars and mountain lions had eaten. This helped biologists learn not only meal preferences but how competition from human hunters might affect felids. Other research had suggested that a decline in deer and peccaries, favorite protein sources for both humans and big cats, might lead to an increase in livestock losses or migration of the cats outside the area. Poaching here, as elsewhere in the region, was commonplace.

I commented on the lawlessness that seemed to prevail in Guatemala and was met with a knowing smile. "Power here does not necessarily reside in the hands of the government," conceded McNab. "There is an incredibly powerful shadow economy in Guatemala based on illegal drug trafficking. This is pumping many dollars into alternative, nonlegitimate uses, such as buying land for farming and timbering." Nonetheless, McNab said that he remained optimistic in the face of persistent institutional corruption, gang violence, anemic government support, and his own harrowing bouts with malaria and dengue fever.

"Sure, I've received death threats," McNab disclosed in a calm voice.

Figure 14.2. Remains of a deer killed and fed upon by a jaguar. The cat follows a pattern different from that of the mountain lion, whose habitat often overlaps with the jaguar. Scientists note that this carcass was not covered and that meat from the head, neck, ribs, and shoulders was consumed, but no abdominal organs and only a small amount of bone. Photo by Carolyn M. Miller

"I've had disturbing late-night phone calls from people I'm sure would love to kill me. Our staffers in the field have been kidnapped and burned out by dangerous characters."

I asked if such animosity frightened him.

"Nah," he said, with a dismissive wave of his left hand. "You keep pushing ahead; you don't let fear stop you. My only real fear is boredom. I want to live fully. But I do worry about people who work for me and for whom I'm responsible. I don't want them to get hurt."

If anything, McNab concluded, his experiences in Guatemala—where governments and nonprofit groups come and go with alacrity—convinced him of the need to persevere. "One can't commit to conserve an area for only two or three years," he emphasized. "When you decide to preserve a place, it has to be forever. Everyone involved needs to know that."

With so many forces militating against success, I asked McNab whether he held hope for the long-term survival of Guatemala's natural wonders. He stared pensively out the window at Lake Petén Itzá, a shimmering lagoon now badly polluted. "I believe that our presence in the forest, doing camera trapping and so on, does make a positive difference in how rural people view their patrimony. They begin to see nature as important—and recognize it as a valuable part of Guatemala." McNab then motioned for me to set down my pen and put away my notebook.

✒✒

The next day I hired Hector, a jovial and garrulous alabaster sculptor from the lakeside town of El Remate, to drive me to the Maya ruin of Yaxhá, an hour east of Flores along a dirt road. My goateed, college-educated companion was that rarity in Petén: a self-described environmentalist.

"Living in balance with the forest is an admirable goal," he volunteered in Spanish, "and it seems to be happening in Uaxactún. But in Guatemala there is a popular saying: '*Siga el pisto*' [Follow the cash]. Politicians and people in our justice system do the bidding of those with economic power. That's our tradition. And conservation, along with everything else, is affected directly."

My driver grew up in Jalapa, in the heart of industrial-strength farm country, and despaired that "it's in my government's interest to keep people uneducated. That way they are easier to control and manipulate." A better future for nations such as Guatemala will come "through educating children and getting them to change their approach, their attitudes, and their habits." Otherwise, he said, "things like saving the environment are simply not part of their consciousness."

While Hector chain-smoked cigarettes and shared wild nuts with the park's caretakers, I wandered alone among Yaxhá's mist-shrouded stone monuments. Although I knew jaguars had been seen occasionally in the ruined city, they were said to have disappeared after the 2005 taping of a popular U.S. reality CBS-TV show, *Survivor*. A crew numbering in the hundreds had overrun the site, and a busy production center provisioned by helicopter had been built nearby to house and feed workers. According to Hector, an on-call driver, the complex included an all-night disco, full bar, and gambling hall. While *Survivor* may have stressed the ability of a few individuals to master a wilderness environment, the show's support staff endured few privations.

There were many birds, monkeys, agoutis, and other small animals at Yaxhá on the morning of my visit, but no evidence of cats. The inanimate exception was a priest wearing a jaguar-skin cape and headdress, depicted in a mural at the site's small museum. The *museo* overlooked the city's namesake, Lake Yaxhá, which translates as "green lagoon." (The water is actually blue, but the Maya are said to have considered blue to be a shade of green rather than a separate color.)

The five hundred–structure archaeological site is less than nineteen miles southeast of famous Tikal but shares a unique relationship with two other cities—Nakum and Naranjo—that form a triangle with Yaxhá amid a network of lagoons, causeways, and low hills. This necklace of navigable water allowed the ancient Maya to move food, goods, and armies easily through the dense jungle. From the top of Yaxhá's tallest pyramid one can clearly see Topoxté, a lakeside ruin that was once a thriving trade center. Unlike at Tikal, many of the abandoned buildings and monuments in these locales are still covered by dense vegetation. Westerners were unaware of Yaxhá's existence until 1904, even though its extensive farms are now believed to have fed tens of thousands of Maya. Without this bounty, made possible by abundant irrigation supplies, Tikal probably never would have become a great regional power.

I asked several of the machete-wielding groundskeepers whether they had seen any recent evidence of *el tigre*. Most shook their heads, but one worker scratched his chin before replying.

"I hear them sometimes at night," he said, eyebrows rising and eyes lighting up. "There are tracks, too, on the edges of the park. I think they come to eat our monkeys and coatis." The man bent down and resumed the perennial task of hand-cutting the grass that grew alongside an acropolis. Then he looked up: "Local people still hunt jaguars for their skins," he said. "The meat is not so good to eat, I think, but you can sell a pelt in the Santa Elena market for seven hundred quetzales [ninety-five dollars]. That's a lot of money for someone like me." The groundskeeper winked before restarting the back-and-forth slash of his machete.

For a twelve-month period, the producer of *Survivor* had agreed to subsidize the admission fees for any schoolchildren who wanted to visit Yaxhá, a boon to local teachers. Two busloads of students arrived as I descended from the site's tallest pyramid, which provided a vertigo-inducing panorama of La Selva Maya. I was pleased for the Guatemalans who had benefited from the TV show's largesse, but I wondered whether jaguars would ever feel safe in returning to a now-busy archaeological park.

Afterward I met over a midday snack with Gabriela de la Hoz, the world-weary Italian expatriate who owned a nearby ecolodge. We sipped cool drinks and compared notes. De la Hoz felt that the biggest threat to wildlife in the Yaxhá area came from illegal settlement near and within the archaeological park, which she believed led to widespread poaching. She gestured in the direction of Yaxhá's lightly patrolled perimeter. "The government routs these squatters from time to time," said de la Hoz, "but they always come back. Settlers live and hunt in Tikal National Park, too."

Hector told me on the drive back to town that he and many other Guatemalans had earned good wages from *Survivor*. "Except for the twenty families that run this country," he laughed, "we *all* need money." The typical annual income for a peasant was nine hundred dollars, Hector said, while a middle-class professional could expect to earn a mere four thousand dollars a year. He had turned his back on his own low-paying field—engineering—in order to live a simpler life sculpting, driving a taxi, and tending to a small vegetable garden with his wife and children.

❦

I dined that evening at a stifling-hot restaurant in Flores. Two fanciful, almost childlike paintings adorned the walls. In one, a happy jaguar family—

mom, dad, and cub—sat calmly next to Lake Petén Itzá. In the other, two serpents coiled around a jaguar poised to attack a man lying prone on the ground, while a Maya shaman danced his magic in the background. Earlier in the day, Hector had refueled his taxi at a Jaguar-brand gas station, which also sold the "Llanta de Pantera" (Tire of the panther). I realized that even though "real" jaguars remained elusive, their stylized, cartoonlike images were constantly evident in the Petén. Late that night visions of felids swirled in my dreams. I caught glimpses of jaguars but could not get near them. They shimmered as if seen through a gauzy veil. Whenever I moved close, the cats moved away, ignoring my entreaties. In one dream, a jaguar studied me with a diffident expression, then turned and walked away, acting as if he had more important things to do.

I woke up feeling frustrated and strangely incomplete. At sunrise I boarded a decrepit bus bound for Belize, which broke down only a few hundred yards from the frontier at Melchor de Mencos. The driver scolded the steaming engine as if it were a petulant toddler. When it refused to cooperate, I joined other passengers in pushing the bus down a straightaway, but the worn clutch would not engage. I grabbed my luggage and, after showing my passport and paying my exit fee, walked across the border into Central America's only English-speaking country—and arguably its last best hope for effective conservation of the jaguar. More than 70 percent of Belize's forest cover was said to remain intact, with as many as one thousand jaguars on the move within its borders.

CHAPTER FIFTEEN

"Living in the Same Place It Always Has"

SEEN FROM THE AIR, THE Cockscomb Basin is a jade carpet of seemingly impenetrable jungle. A backdrop of serrated hills and sharp pinnacles—some rising to nearly 3,700 feet above the nearby Caribbean—gives the watershed its name. The jagged profile of one particularly prominent ridge does, in fact, resemble the wobbly flesh atop a rooster's skull. So thick is this vegetative maze, rent by rivers and ravines, that when the biologists Alan Rabinowitz and Ben Nottingham used radio transmitters in the 1980s to track collared jaguars, signals were quickly absorbed by the millions of vascular plants. The men were forced to climb precipitous slopes or board airplanes in order to pinpoint cat locations. Not long after its 1984 establishment, thanks to Rabinowitz's prodding, as a 3,900-acre preserve, the watershed was reclassified as a wildlife sanctuary. It was expanded to 102,000 acres in 1990.

The Cockscomb was lightly populated even during the glory days of the Maya civilization, about a thousand years ago. Though not technically a rain forest—it needs more year-round moisture to qualify, and is instead labeled a "moist tropical forest"—the basin is an ideal environment for many species. As one of the least disturbed and most accessible protected ecosystems in Central America, it is a coveted destination for wildlife experts. Creatures now scarce elsewhere flourish here. The roster includes tapir, peccary, deer, howler monkey, iguana, curassow, and all five regional cat species: jaguar, mountain lion, ocelot, jaguarundi, and margay. Every carnivore seems to find ample prey. Springs deliver year-round water, and old logging roads have become animal highways, particularly at night. Some scientists believe the entire region might hold as many as two hundred jaguars, potentially the highest small-area concentration in the world.

In April 2006 I took a twilight hike about two miles up the dirt road toward the reserve from Maya Center. The population of this small community of Maya includes families forced to relocate from Cockscomb when the park opened two decades earlier. In return, family members had received assurances that some would be hired as guides, wardens, and managers.

As I left the village, I soon noticed that the entire forest along one side of the road had been chopped down since a visit five years earlier. High bush had morphed into an orderly citrus orchard. Beyond orange trees, the gash of a rock quarry was visible on a nearby hillside. Even at dusk, large trucks full of crushed rock roared up and down the bumpy road. Gazing past the quarry, I could make out newly cleared areas where farmers and ranchers had wrestled the jungle into agricultural submission.

Silence returned to the forest between the jarring passage of each dump truck. I followed a narrow lane that forked to the left and crossed a creek, where a sign indicated that I was six miles from sanctuary headquarters. The track plunged immediately into tall second-growth forest. It was cool and peaceful in this soft cocoon of green, gauzy light. At the bottom of a steep slope, where the canopy merged above my head, I was bathed in the blinding flash of a camera trap. I felt a fresh pump of adrenaline as I realized that once again I had entered the domain of the jaguar. Big cats—not humans—ruled this environment. The Cockscomb was their world, not ours. (A researcher later confirmed that jaguars were photographed regularly here and at two other camera sites along the entrance road.)

A bolt of anxiety rattled me. I wondered how I would react if a jaguar or mountain lion appeared in the fast-fading light. This was prime hunting time for such felids. My intent would be to stand my ground, look as large as possible, and speak in a loud voice. But would I actually *do* this? I was reminded of an afternoon in the Santa Monica Mountains, near Los Angeles, when I inadvertently roused a well-concealed mountain lion from its day bed and found myself backing down the slope of a treacherously steep canyon. My luck held when the lion quickly lost interest and vanished.

On this evening I wished for a walking stick, though it would be of little use in an attack. The defense it provided would be mainly psychological. Not finding a suitable piece of wood, I kept to the middle of the road and looked back every few minutes to see whether I was being followed. Walking alone near dark in this jaguar sanctuary made me jumpy.

I encountered no large animals but found one smallish set of cat tracks in the soft earth, probably left hours earlier by an ocelot. There was other native wildlife to behold, though: a slithering vine snake, a saucer-sized toad, bats, lightning bugs, several yellow parrots, and a Jesus Christ lizard (so named for its ability to stride rapidly across the surface of water). Rounding a turn, I roused three large birds—curassows or crested guans—perched above me. Both pheasantlike game species are threatened and neither is easily found in Central America outside of parks.

Darkness and mosquitoes engulfed the forest at last, so I turned and walked swiftly along the road back to my room. I was staying at a leafy compound overseen by Ernesto Saqui, the Cockscomb director from 1991 to 2004, and his artist wife, Aurora García. Saqui, a heavyset Mopan Maya and former schoolteacher, had grown up farther south, near Punta Gorda, and had become an astute environmental educator.

"Conservation is important, not just to protect jaguars but to protect people," Ernesto Saqui told the author Donald Schueler in *The Temple of the Jaguar.* "However, this is a new idea for us. We are surrounded by communities that used to fish and hunt [in the Cockscomb] and it is important they understand why the sanctuary is here." Some people were bitter about being forced to leave their forest home, he told Schueler, and it helped that they were given an important stake in its future.

Ernesto had resigned from Cockscomb when Aurora developed health problems and become too ill to oversee their various businesses and young children. Ernesto pitched in while his wife recovered and also served as chief of the village council. The Saquis extended to me a warm welcome under trying circumstances. Tourism was down in the Cockscomb this season. It was clear that Maya Center residents were benefiting from visitors, though not at the level of prosperity enjoyed in upscale resorts along the coast, where most visitors preferred to stay. This spoke to a universal complaint in developing countries about the fickleness of service-based economies. While a tourism infrastructure had been created in Maya Center, there was no guarantee that foreigners would arrive on a regular basis. Residents were barely getting by.

After getting reacquainted with García in the lodge's dining room, I was approached by an intense-looking man who identified himself as Sondro. A Maya whose home was near the Guatemala border, Sondro told me that he was a fine-art painter consigning canvases for sale at local shops. After I described my mission, Sondro confessed that he, like most people living in jaguar country, had never seen such a cat in the flesh. A moment later the artist unfurled a primitive, Henri Rousseau–like canvas of an idyllic jungle scene, rendered in brilliant acrylics. Surrounding an idealized, side-view jaguar were macaws, toucans, orchids, and tropical fruit trees. This DayGlo cat was clearly king of the forest.

Soft-spoken and modest, Sondro said that he had three boys and three girls to feed back home. I assured him that his works were lovely, but said I would not be buying any. Instead, I treated Sondro to lunch, over which we talked about the animals he admired. "Jaguars are the most powerful," the artist declared. "They're like gods. I am glad a few are still among us Maya, even if we don't see them."

The next morning at breakfast I joined a Canadian couple who had taken a "night hike" the previous evening, led by a local guide named Julio. Their most memorable experience had been to smell and hear a herd of white-lipped peccaries. These piglike animals—called *warries* or *waris* in Belize—are legendary in their unpredictability. Angry peccaries have been known to chase and form threatening circles around predators, including jaguars. Stories circulate about big cats being chased into trees and kept cornered there for days. Such "strength-in-numbers" tactics are used effectively by a number of herd animals in order to avoid decimation. In Africa, zebras and Cape buffalo employ a similar strategy to stave off attacks by lions and leopards.

As the Canadians talked, Saqui flitted in and out of the restaurant, engaged laughingly with García and their children. I caught up with him over dessert, and we talked about jaguars roving outside the sanctuary. "These days they are staying away from our village," Saqui told me. "The cats in this area seem to remember that we had to kill a jaguar a few years ago after it began taking dogs and chickens. Still, we sometimes see jaguar tracks nearby, as close as a few hundred feet from our home."

Saqui invited me into a part of the family compound that faced the forest, cool in the dappled light of midday. I followed him along a trail marked with signs in English, Spanish, and Mopan Maya. They labeled medicinal flora that his wife used in her practice as a *curandera,* or natural healer. As a niece of the renowned Eligio Panti, who died at 103 after nearly a lifetime as Belize's foremost "medicine man," García was admired nationwide for her extensive knowledge of the endemic plant pharmacy.

"Only rarely does anyone see a jaguar anymore," Saqui told me, as we walked beneath the emerald canopy. "They are very solitary, secretive cats, yet it's important to come to where jaguars live, isn't it? At least here you can see their tracks on the ground, their scat and their scrapes. At least you know something so sacred is still alive, living in the same place it always has. Amazing, no?"

On a Tuesday morning, muggy and cloudless, I strapped on my backpack and trudged the seven miles through low foothills from Maya Center to the Cockscomb Visitor Center. I was passed by vans full of eager-looking foreigners, the financial lifeblood of this refuge. Most were day-trippers: white, affluent, English-speaking. During this week I would be the sole guest in a dormitory built to accommodate overnight visitors.

Despite the oppressive heat and cautionary words from the warden who checked me in, I was itching to take a look around. I promptly set off on a six-mile scramble up the Tiger Fern Trail, which wound its way through thick second-growth bush before zigzagging to the top of a ridge. (An estimated 90 percent of Cockscomb's overstory trees were destroyed by Hurricane Hattie in 1961, while many of its remaining hardwoods were removed during timber operations that ended in the mid-1980s.) I caught my breath and mopped my brow at a picturesque campsite set beneath towering pines. The view to the east encompassed the entire steamy basin, with Victoria Peak looming through the haze. After wading through neck-high ferns and brambles, I descended into a verdant canyon on the backside of the hill. Between hairpin turns I spooked a deer and a curassow before arriving at a spectacular double waterfall. After many hours of trekking, it felt delicious to strip naked and swim in the cool, deep plunge beneath the falls. I let the cascade pummel my vertebrae with a restorative massage.

The setting was idyllic and I was tempted to linger, but sunset would arrive soon. I headed to the dorm, backtracking upslope and through the scratchy thicket of bracken and thistle. By the time I reached my room I was dizzy and red-faced. I had come to Cockscomb in April, the hottest and driest month of the year.

At twilight I had a knock on my door from Rebecca "Becci" Foster, a doctoral candidate in jaguar ecology from England's University of Southampton. The Wildlife Conservation Society, an enduring sanctuary presence, had funded much of Foster's fieldwork. As Cockscomb's then-senior researcher, the scientist had studied human-jaguar conflict along the perimeter of the basin for more than two years, following the 2002 departure of a husband-and-wife WCS team, Scott Silver and Linde Ostro.

Blonde, thirtyish, and exceptionally fit, Foster was uncommonly focused and informative. Her inquisitive mind and darting eyes seemed to miss nothing. She had spent much of her three-year residency in Belize with her fiancé, coresearcher Bart Harmsen, also a Ph.D. candidate in jaguar ecology from Southampton (and home in England at the time of my visit).

Foster and I immediately began swapping anecdotes about large felids. She was surprised that the driver of the truck that had driven me from Dangriga to Maya Center had described seeing jaguars—both alive and dead—during his weekday trips along the Southern Highway. The truck driver had expressed concern that increased traffic was killing or injuring many animals that had previously been able to cross this principal north-south artery.

Foster explained that a field researcher's work is never done. Enormous amounts of time are devoted to the tedious and unending task of setting up and checking camera traps, followed by the recording of such data as scat counts and photography yields. These chores are considered essential, particularly when area jaguars are not fitted with radio collars. Within Foster's study zone at the time were forty-six double-sided traps, for a total of ninety-two cameras. Nineteen traps were within the sanctuary and twenty-seven outside. Based on data collected between 2001 and 2007, Bart Harmsen estimated that about forty individual jaguars lived in the basin, or about one per four square miles. This was considered an unusually high density, with most jaguars concentrated in low-lying areas. A large number were documented just outside the park, too, but ranchers and farmers were killing these cats off at a rapid rate.

"Inside the sanctuary the ratio of males to females seems to be about two to one," said Foster. "It can be more like nine to one for shorter periods, such as we find in some of our one- to two-month surveys." The continuing appearance of new individuals suggests that the Cockscomb population has bred well.

I told Foster that cameras targeting jaguars along the U.S.-Mexico border occasionally took pictures of drug smugglers, undocumented immigrants, hunters, and even nude hikers. She countered with a story of four bare bottoms mooning one of Harmsen's cameras. Another photo captured trekkers carrying an ironing board and steam iron through Cockscomb toward Victoria Peak. The prank was part of an international movement of "extreme ironing" in the world's most remote places.

Foster was house-sitting for a couple of weeks at the home of a Dutch expatriate who lived just outside Maya Center. Her duties included the care of a young ocelot, for which the Dutchman had paid one thousand dollars. The foreigner had made the purchase from a hunter after being told that the animal's mother had been killed, leaving the cub unable to fend for itself. The adopted animal wore a collar and wandered near the expat's home during the day, coming in at night for a meal and a cuddle. I soon learned that sales offers of orphan cats are common, despite their illegality. The Belize Department of Forestry had granted the Dutchman a permit because the orphaned ocelot had nowhere else to go. (Weeks later, in Panama, a man told me that he recently had been invited to purchase a jaguar cub for fifteen thousand dollars.)

Unsurprised by my report of widespread hunting of jaguars to the north, Foster noted, "Such killings are happening all over Belize. It is bound to increase as pastures expand. Development—including the conversion of forest to other uses—is the biggest threat to all wildlife here." She noted that both the mouth of the Sittee River and the adjacent village of Hopkins, less than twenty miles from where we stood, were filling with modern condominiums and homes, most built or bought by foreigners for investment or retirement. Along stretches of the recently paved Southern Highway, nearly pristine forest was being replaced daily by farms, fruit plantations, and cattle pastures.

"Lethal control of jaguars here is not illegal if the cat is threatening life or livelihood," Foster explained, "but it must be reported to the government. In effect, the government owns the jaguar carcass. I would not necessarily want or expect these [jaguar killers] to be prosecuted, especially if the cat was causing continuous trouble around a village. [But] I find it frustrating when people who can actually afford to manage their cattle better do not take sensible measures to protect their livestock and resort to shooting jaguars instead." She noted that the Forestry Department, which investigates reports of depredation, had a dedicated staff with limited resources, including a scarcity of vehicles and fuel. (At least one large farm near the sanctuary later shifted to more jaguar-friendly management practices and Forestry subsequently allocated more personnel to the resolution of human-jaguar conflicts.) Foster and I agreed to meet the following day, after I had done more exploring on my own.

Figure 15.1. Drag marks left by a 250-pound calf pulled under a fence and across a road by an adult male jaguar in northwestern Belize. The cat, a research subject, was tracked and killed later the same day. Photo by Carolyn M. Miller

It was a full-moon night, a time when I often have trouble sleeping. Whether this is a result of the night sky's glare or heightened activity in the natural world, I am unsure. But on this occasion the Cockscomb air was still like death and as damp as a washrag. I lay on the thin mattress, too hot even to be covered by a sheet, and could not stay in any position longer than a few minutes. My mind raced in a million directions and I finally gave up all attempts to sleep. I arose around four o'clock and, not bothering in the steamy heat to don clothes, walked outside onto a wooden deck that faced the jungle. Strands of mist clung to the canopy and glowed silver in an exquisite display of lunar light. I peered into a stand of liana-wrapped trees crowding their way into cropped grass at the edge of the compound. A chacalaca, one of the most talkative birds in the Belize forest, gave its melodious call. "Could this feathered fellow," I wondered, "be under the mistaken impression that dawn is close at hand?"

Only a few seconds had passed when suddenly, from beyond the rusty capture cages Rabinowitz had used twenty years earlier, came a low, hoarse rumble. It resembled the discharge of a punctured exhaust pipe: *Uh-uh-uh-UH!* After a pause during which even the chirping of insects cut off, the guttural grunting resumed, starting soft and building to a crescendo: *Uh-uh-uh-UH!* This was a raw sound, commanding and primal. More cough than roar, it echoed through the trees a third time: *Uh-uh-uh-UH!* The fine hairs on the back of my neck stood up. This was a series of feral, intimate woofs, loose and uninhibited. "Oh my God," I thought, "that's a jaguar! I've got to get dressed and take a look; this may be my only chance to see one."

But my spellbound body balked at its brain's command. "Better stay right here," a prudent interior voice advised. "This cat may be on the hunt and might not appreciate being interrupted." Even if it meant me no harm, I knew the jaguar would see me before I saw it—assuming I could make out its form in cloaking shadows. Becci Foster's confident assurance a few hours earlier that wild jaguars are never aggressive toward nonhostile humans vanished like a popped bubble. No matter what she had said, I was not eager to confront such a fearsome predator unarmed in the wee hours. For the second time in as many days, I felt ill prepared to fulfill my ultimate fantasy. Running off to meet an apex predator in the dark demanded more than curiosity and desire.

As I weighed this discouraging reality, still itching to follow the craggy roar to its source, an intense rain shower swept down and drenched the Cockscomb, shuttering the moon behind thick clouds. The cat stopped vocalizing and was not heard again.

I crawled back into bed and lay awake, every few seconds replaying the sound of the jaguar, like a musician's endless tape loop. The tantalizing snippet would not stop. My sense of engagement in the moment was like an addict's high, sending me on waves of giddy exhilaration and filling me with a desperate longing to see this cat before it slipped away. I willed the animal with my entire being to move into the moonlit compound, where I could observe it easily—and safely. No dice.

I finally clicked on a light and recorded thoughts in my journal. "I can't believe it," I wrote. "What I heard a few minutes ago was exactly the same sound I've heard in zoos and TV documentaries. It is a low vibration, gruff and breathy. This call surely crosses the species barrier, since every animal wants to know when a top predator is near. Was the jaguar gathering its cubs? Looking for a mate? Warning intruders? Or merely advertising its presence? I wish I knew, but what a thrill to hear such an assertive roar on a jaguar's home turf." My excitement finally crested, and, like a diabetic slipping into a sugar-induced coma, I fell into fitful slumber.

At seven A.M. I was seated on the steps of the deck drinking ginger tea when a red brocket deer—a delicate, knee-high ungulate that is a favored prey for large cats—stepped cautiously out of the forest and tiptoed in my direction. The doe browsed dewy leaves and, after noticing me, strolled casually behind a curtain of foliage. The diminutive species, called antelope by Belizeans, is almost completely gone from unprotected terrain. I felt lucky to see this deer so unafraid. Obviously, humans had never hunted the dainty creature.

The incident inspired me to take a two-mile hike along the Antelope Trail, which rambled to the edge of the basin over a series of low escarpments and shallow streams. Other than a troupe of howler monkeys, my only visible companions were several species of multihued birds: parrots, parakeets, toucans, curassows, guans, and trogons. The jungle was alive with color and movement, but it had swallowed my early-morning visitor without leaving a trace.

Upon my return a cheery Cockscomb warden confirmed my predawn experience. What had happened was not a dream. The man had discovered fresh jaguar tracks on the Wari Loop, a trail less than a quarter-mile from my room. On the same path earlier in the week, the warden said, visitors had encountered a tapir. I had seen the large, three-toed dinosaurish footprints of such an animal—perhaps the same one—on the Loop the previous day. Deer and peccary tracks had also been much in evidence, and I had heard an armadillo scuttling through the underbrush. "These Cockscomb jaguars are well fed," the warden deadpanned. "They go where the food is." I smiled and breathlessly recounted my story.

Late that afternoon Becci Foster returned from her fieldwork and extended an offer I could not refuse. She invited me to accompany her the following morning on a survey during which she and the WCS guide Emiliano Pop would check camera traps along the Sittee River, a few miles beyond the park's boundary. This would be my first chance to watch Central American jaguar researchers in action.

CHAPTER SIXTEEN

"Pretty Well Hunted Out"

MY DAY BEGAN AT 6:30 A.M., when I drove from the research station to the Dutchman's ranch. There I met Becci Foster and Bud, the marvelous but needy ocelot. The scientist stowed her gear in preparation for the short drive to Emiliano Pop's home nearby. The passenger door of Foster's vehicle bore footprints from Bud's paws. It seemed that this anxious cat—perhaps made hopelessly neurotic through early separation from its mother—had a habit of trying to jump into the cab whenever his guardian prepared to drive away. We discouraged lonely Bud by rolling up our windows and edging out of the driveway.

Ocelots have spotted tawny fur somewhat like a jaguar, but are considerably smaller, about three feet in length with a weight of twenty to thirty pounds. Distributed over much the same area as the jaguar, plus the southernmost bit of Texas, the solitary ocelot is a mostly nocturnal carnivore, preying upon birds and other small animals. Like other cats, it is sometimes killed and eaten by jaguars. Hunting ocelots is prohibited in seventeen of the twenty-one countries where the species occurs; nonetheless, the cats remain a popular sale item on the black market as pets and for their attractive pelts.

We soon pulled up to a smiling Maya man who was sharpening a machete as he stood next to his beaming wife and giggling, wide-eyed children. The Pops seemed the epitome of the phrase "one happy family." Their home was modest, surrounded by a large yard filled with fruit trees. Like their neighbors, the Pops were fluent in English, Spanish, and a local Maya dialect.

Emiliano Pop was a respected, forest-savvy WCS escort who had tracked howler monkeys—successfully reintroduced into the Cockscomb after hurricanes and disease had wiped them out—in addition to many jaguars. After climbing into the pickup, he directed us through orderly rows of orange trees to a barely visible pathway leading toward the Sittee River. The waterway flowed sinuously through a broad, forested plain. Since 1990 much of its fertile bottomland had been converted to plantations of citrus and banana.

By eight o'clock we were on a swift march along a trail that swerved around innumerable tree roots, bamboo stands, and thorn-pronged shrubs. The thick air was oppressive and damp, causing my eyeglasses to fog repeatedly. An overcast sky spit intermittent rain, keeping the narrow track mucky and slippery. After about forty-five minutes we arrived at our first site, its twin cameras strapped to trunks facing one another. Pop motioned us to halt while he sidled at a low angle to disable the infrared trigger mechanisms. (Researchers must be careful when checking camera traps to avoid taking pictures of themselves: body heat and movement will trip shutters that are approached improperly.)

My companions followed a rote procedure. They checked film and batteries, replacing components as needed. During the preceding ten days only a few frames had been snapped, so this film would be collected later. (Digital technology was slow in coming to camera traps, and film canisters had to be sent to the nearest city for processing and printing.) Foster placed fresh desiccants in the security boxes that held cameras and their flash units. Desiccants lower humidity in the boxes, protecting circuits and preventing film from becoming stuck.

We stopped at two more installations, each about forty-five minutes from the last, before retracing our route back to the orchard. We occasionally sipped water from plastic bottles, pulled snacks from our rucksacks, and stopped to examine jaguar scrapes—patches the size of dinner plates nearly devoid of forest litter. I would have walked past them if not for the expertise of Foster and Pop. This is where cats had sprayed from their urethra and anal glands. Foster also showed me an otherwise smooth, foot-wide trunk that bore deep jaguar claw marks, probably saturated with distinctive

Figure 16.1. The ecologist Rebecca "Becci" Foster of England's University of Southampton conducts field research on jaguars along the Sittee River in southern Belize, April 2006. Field assistant Emiliano Pop, a Belizean Maya, is at left. Photo © Richard Mahler

oils from glands around the cat's toe pads. The scrapes and scratches were positive signs that suggested the presence of more than one jaguar.

"In the Cockscomb," said Foster, "we may find as many as one hundred fresh jaguar scrapes along an eight-mile trail over a one-week period. We suspect this happens when a female is in heat, prompting males to spray and scrape while competing with one another for her attention." Exactly how jaguars communicate through such odors is unclear, though its importance is undeniable. All felines, including domestic housecats, use finely tuned sensory receptors in their mouths and noses to "taste" and evaluate smells left by other cats.

Pop set a brisk pace, followed by Foster, then me—nearly thirty years her senior—gasping for breath. There was little stopping, no dawdling, and minimal conversation. It was all I could do to navigate the uneven path through blurry spectacles. (I had nothing dry to wipe them on.) Wispy clouds swept above us, shifting levels of brightness beneath the canopy like an electric dimmer switch. I stumbled on, dealing with an unending succession of branches, leaves, twigs, roots, streams, puddles, mud, brambles, vines, ferns, and boulders. I felt new admiration for field biologists, knowing I had far to go before I could meet the physical demands of their research projects.

We continued parallel to the Sittee, glimpsed infrequently through dense foliage. My initial elation at being in the bush sank as I realized we were likely to encounter few animals. Our surroundings seemed too quiet, too still. It felt as if the wildlife had vanished, leaving a ghost forest behind. When I remarked upon this, Pop shrugged: "The area has been pretty well hunted out." Although it was not much closer to a village than was the Cockscomb, the Sittee ecosystem had no government protection. As a result the game species, for which humans competed with wild carnivores, had been killed or forced into hiding.

As Foster drove us back to Pop's home, Emiliano recalled that a jaguar —the same one described to me by Ernesto Saqui—had hung around Maya Center in 2004, devouring canines and fowl late at night. "Dogs were taken right from doorsteps," he marveled. "We had no choice but to shoot that cat. We didn't want to, because the Maya still hold great respect for jaguars." His comments resonated with those I had heard from other rural residents in Belize, Guatemala, Honduras, and Mexico. Some measure of admiration for jaguars seemed universal in the backcountry. Pop recalled that before the marauding cat's death some in Maya Center had worried that it might come after their children.

"Have you ever heard of a jaguar actually killing a person?" I asked.

"Yes," Pop replied. In the late 1980s, he said, a story circulated about two Maya boys in the Toledo District, near the Guatemala frontier. The youngsters had gone into dense forest to hunt and fish, but apparently became lost and were stranded outdoors overnight.

"Those poor kids never returned," murmured Pop. "All that was found was their torn and bloody clothes." Local Maya claimed that cat spoor was found not far from the tattered garments. The assumption was that a jaguar had killed and eaten the children, but no remains were ever found and the allegation could not be proved.

In other tropic locales I had heard stories about men and women being followed for miles by jaguars that had never threatened them. While in the Cockscomb, Alan Rabinowitz several times discovered that a jaguar he was tracking had circled back and followed him, apparently out of curiosity. On one occasion, the researcher became suspicious and turned around, confronting a jaguar that stood—then sat—only fifteen feet behind him. The cat, which was not aggressive and apparently had trailed Rabinowitz for some time, eventually got up and walked calmly into the forest. The Costa Rican researcher Eduardo Carrillo has reported similar incidents. It is as though the curious cats were checking up on human interlopers, then gently escorting them out of their home territories. Such behavior suggests that at least some jaguars neither harbor fear of humans nor view them as prey animals.

Pop's reverie prompted Foster to reveal that she kept a running list of persistent, unproven beliefs about jaguars that circulated locally. "For instance," she began, "some Belizeans keep insisting they've seen jaguars with stripes, like tigers, rather than the usual rosettes and spots. Others claim to have seen black jaguars, which are considered rare in Central America and southern Mexico." (I later heard conflicting opinions on the latter subject, with the expert consensus being that while at least a few black specimens have been documented in Belize and as far north as Mexico's Chiapas, the vast majority of melanistic jaguars occur in South America.)

Speculation also was rampant about how jaguars move large prey animals after killing them. On occasion they take down 350-pound tapirs and 1,000-pound cows as well as full-grown horses and mules. In South America—where the largest male jaguars rarely exceed 300 pounds—the cats are said to move carcasses weighing nearly a ton. Some believe that jaguars hoist these dead animals onto their own backs, while others think carcasses simply are dragged along the ground. The only demonstrable fact is that jaguars somehow transport bodies far heavier and bigger than their own from open pastures into sheltering forests. The exact logistics of these daunting performances are unclear, although scientists know that all cats have unusually dense muscle fibers.

I next sought information about past and present big-game hunting in Belize, but Foster had little to offer. Several years earlier I had interviewed Emory King, a legendary figure in contemporary Belize history, about high-priced jaguar safaris during the colonial era. King, a U.S. sailor who during his twenties virtually washed up on the shores of what was then British Honduras, described a lavish aerie in the Mountain Pine Ridge that had been popular with hunters and well-off expatriates. Besides guides and tracking dogs, a visitor to the remote lodge could find liquor, prostitutes, and gambling. The property boasted its own airstrip and could otherwise be reached only via a long, circuitous dirt road. The dilapidated facility was purchased in the 1990s by the movie director Francis Ford Coppola, who reopened Blancaneaux Lodge as an upscale hotel and fashionable restaurant. "Francis comes here every April to celebrate his birthday," a spokeswoman told me. When I asked her to confirm the colorful history of the place, as detailed by Emory King, now deceased, the woman only blushed and rolled her eyes.

Toward the end of a long day of checking camera traps and jaguar signs, Foster and I replaced a trap near Maya Center that had been vandalized—its camera stolen and chain cut. The researcher had decided to set up a new trap in a more hidden location on the south side of Cabbage Haul Creek, about two hundred yards from the village.

Even this simple task took a surprising amount of time. Earlier in the week Foster had obtained permission from a property owner to trespass on his land in order to access the new survey location. Fording the creek on stepping-stones, we used mobile GPS units to get coordinates, which needed to be close enough to the previous camera not to compromise the study's integrity. We walked through backyards, scattering chickens and rousing dogs. Finally, we followed an overgrown footpath into the bush.

At our first preferred location the smooth base of a cohune palm proved too thick to secure an antitheft chain. We located a smaller-diameter trunk a few yards away. I held the camera against the tree while Foster crept on hands and knees to determine whether the electronic eye would be triggered properly. It was.

Next, Foster whacked surrounding vegetation with her machete to provide a clear view across the path at knee height. We strapped a metal frame to the trunk with thick wire before double-checking film and batteries. A separate activation unit and set of batteries had to be tested as well. The box in which the camera sat was examined for proper padding and desiccant, then clamped shut and secured with a chain.

Working as a team, Foster and I activated the camera one last time to make sure it was taking pictures. She attached a hand-lettered waterproof sign that explained the setup's purpose in simple Spanish and English. Researchers held hope that if their scientific goals were understood, vandalism and theft might decrease. Cameras were broken or stolen with regularity here, as in the American Southwest, costing the project thousands of dollars. Finally, Foster logged the trap's coordinates into her notebook. Throughout this procedure, we swatted biting flies and swished away mosquitoes while trying simultaneously to keep dripping perspiration out of our eyeglasses.

As we drove back into the park, the promise of a cold beer had never sounded so good. Guiding the pickup through the forest, Foster explained that her ultimate goal was to determine how jaguars use habitat that is becoming increasingly dominated by people, and when, where, and how domestic animals were being lost to jaguars—and what to do about it. This problem did not lend itself to easy remedies.

"You can't blame the cats for going after livestock, which to them is just food," Foster emphasized. "Ranchers ask, 'Why not fence the Cockscomb to keep its jaguars inside?' Well, to begin with, it's impractical and unwise to put a high fence around such a park. The bigger problem is that most jaguars going after livestock are not living within the sanctuary, but outside its borders instead. They've always been there, they were simply not noticed in the past because ranching was never done so close to the park boundary."

There was frustration in Foster's voice as she described a cow-killing jaguar reported to her twenty-four hours earlier.

"It's taken four calves since Thursday."

"Wow," I said, "that's only one week ago."

"No, I was told that yesterday and the attacks had already happened by then, so this cat took that many calves in four or five days."

"Not good, is it?"

"No, it isn't."

Like other large carnivores, some jaguars on occasion will kill more animals than they appear to need for survival, leaving large parts of a carcass for vultures and other scavengers. When this happens, ranchers are understandably annoyed.

In a subsequent e-mail interview, the researcher Bart Harmsen noted that the lowland part of the Cockscomb "is excellent agricultural land" coveted by many ranchers who "suggest the jaguars can have the highlands." But the cats *also* prefer the comparatively flat, low-lying areas, which harbor more prey. Harmsen said his biggest fear was that a government official someday might simply cut off a part of the sanctuary and sell it to wealthy buyers, perhaps those offering tempting bribes. If that happened, Belize would be following the lead of some less stable and more corrupt countries in the region, where protection of designated parks is more concept than reality.

❧❧

While scientific research is essential, some of the best information about the practicalities of jaguar conservation comes from men like Nicasio Coc, whose training is informal. Late one afternoon I sat down with the director of the Cockscomb Basin Wildlife Sanctuary, a Kekchí Maya born and raised in southern Belize. A slender man in his late thirties, Coc was unassuming yet forceful in his opinions.

"Poaching is a big problem and getting bigger," Coc sighed, after we joined Becci Foster and a colleague, former New York City zookeeper and Columbia University graduate student Ferdie Yau, on the screened-in porch of the research team's bungalow. Yau was midway through a four-month stint in the Cockscomb.

"Hunters sneak onto our land," Coc said, with no trace of anger. "But the Belize Audubon Society, which manages Cockscomb, doesn't want us to carry guns or make arrests, only to give warnings. Each game warden carries only a map, machete, GPS unit, and two-way radio." (In contrast, well-armed wardens at some game parks in Africa and Asia are instructed, as a strong deterrent, to kill poachers on sight.) Coc said that poachers and other intruders are reported to the Department of Forestry, the agency charged with carrying out enforcement of wildlife laws. Citations are infrequent, according to the director, and rarer still are fines or jail terms. On occasion, soldiers from the Belize Defense Forces also patrol Cockscomb.

"The soldiers do this during big holidays like Easter," said Coc. "We know poachers enter the sanctuary to kill game for special meals, so we need extra help."

I asked what species are targeted. Coc ticked off peccary, gibnut (a small rodent), guan, curassow, deer, scarlet macaw, and agouti. Some harvesting is done to fulfill Maya traditions, he noted, but other hunters consider it their long-standing right to take game wherever they find it. Wild animals, for some people, constitute a resource free for the taking. The problem for the Cockscomb jaguars is that except for the macaw, every creature on Coc's poaching list is an important food source for the cats. When such prey become scarce, hungry and desperate jaguars might kill and eat domestic animals, from dogs and chickens to calves and foals.

Coc speculated that many hunters were low-income immigrants from El Salvador, Guatemala, and Honduras. (This theory was later challenged by one researcher, who contended that most newcomers cannot afford guns, bullets, or hunting licenses.) A high percentage of Belize's Central American immigrants work on plantations for minuscule wages. Coc maintained that many of these workers grew up without a conservation ethic or a tradition of respect for park boundaries, property rights, and government authority in general. More to the point, they were inclined to behave as their fathers, mothers, and other relations had for generations.

"Immigrants are surprised at how much wildlife remains in Belize," Coc said. "Many see this as a gift from God. For extremely poor people, hunting is a way to save money that is needed for basic expenses like medicine, clothes, school supplies, and transportation."

Environmental education is a challenging issue for park managers and wildlife scientists everywhere. The buffer communities outside Cockscomb receive special attention from its director, but progress is slow. "Older people often don't understand what we're doing," said Coc, "and our explanations may be too scholarly for young children. But old heads are so hard, maybe it's best to start with kids." He was gratified to see more young visitors brought each year by local schools.

"As for the parents," said Coc, "I explain to them that this is not just a park for animals, but to protect our watershed, to provide clean air, and to preserve plant diversity. Some respond by saying, 'We need money [from natural resources] now, not later.'" Coc said that it was difficult to get his own father to understand Cockscomb's long-term value: "'Protect the *what?*' my dad will ask. 'The *watershed?* What is that? To keep the water clean? It already *is* clean.'" The park director lamented that some elderly Belizeans "can't get their minds around such ideas. It reflects their limited schooling and the strength of their traditions." Still, Coc was optimistic about the future of Cockscomb's alpha predator. "I don't believe many in our communities are killing jaguars," he insisted. "I think if the 'tiger' were no longer in Belize, well, the people really wouldn't like that."

After Coc left the cabin, I resumed my conversation with Foster and Yau. They struck me as highly capable people who had the choice of many career paths. I asked point-blank: "What interests you about jaguars? Why go to such great lengths and endure so many hardships in order to study them?"

Foster struck a reflective pose and answered first. "It's partly because they're so incredibly secretive," she said. "Here is a big animal that can and does live all around us, yet is hardly ever seen. It is smart as well as mysterious. We truly know very little about it." She gazed into the middle distance for a moment. "Besides all that, its jaws are amazingly strong and it is an extremely effective predator, but for some reason chooses not to attack people." Foster paused again. "And, of course, a jaguar *is* beautiful."

There was silence for few moments. Night had come quickly, as it always does in the tropics, and the screened porch was aglow with the flickering light of an oil lantern and a candle jammed into an empty wine bottle. Speaking in a meditative tone, Yau expressed admiration for the strength and fearlessness of jaguars. Like Foster, he marveled at the fundamental unknowns about *Panthera onca:* "I ask myself, 'Where are the females and cubs?' We hardly ever see them, yet we know they're out there. And how *do* we resolve human-jaguar conflicts? That's a very pressing issue."

More quiet ensued before my interviewees turned the table. "So," Foster began, "what makes *you* so passionate about jaguars?"

"Yeah," echoed Yau. "Why write about them?"

I closed my eyes and thought for a minute. "Hmm," I mumbled. "There are so many reasons. For one, I relate to them. I like it that jaguars are generally quiet, cautious, self-sufficient—and rather inscrutable. They are observers, like writers, and I admire their resourcefulness and adaptability. But most of all I simply consider them to be a kind of miracle, one of many in nature I don't feel we can afford to lose."

Three pair of eyes focused on the oil lamp's flame and each of us receded into our own thoughts. I felt heartened by this discussion. The musty, decaying bungalow was a place where field scientists and sanctuary managers—including Rabinowitz and Saqui—had compared notes about jaguars over a span of twenty-five years. This old lumber camp was a conservation epicenter. Saving big cats was an enormous undertaking, and I respected the men and women who worked so tirelessly on their behalf, despite enormous odds against long-term success.

"I know we're making progress," Yau whispered, breaking our musing with a sideways glance at Foster. "Whether it's fast enough is an open question; one that can't be answered."

❦

In Belize I found cause for equal measures of hopefulness and despair. Cockscomb director Coc recounted a disturbing story about a trespasser who was apprehended while walking around Cockscomb with a GPS unit. When asked why he was there, the man shrugged his shoulders and told a game warden that he had been assured of receiving a permit for logging in half of the Cockscomb, where commercial tree cutting has been banned since the 1980s. He was not forthcoming about who in government offered such assurance.

"The status of the park can be revoked by ministerial word without any kind of vote," confirmed Bart Harmsen, coordinator of research in the Cockscomb sanctuary during 2007–2008. The sanctuary "is always in danger. There have been times when we know people have been trying [to reduce its size], but so far there has not been a minister who reacted [positively] to this suggestion."

Such testimonies underscore the insecure status of national parks not only in Belize—with an impressive ninety-seven individual protected areas—but in many other countries as well.

I left Cockscomb early the next morning, hitching a ride with Julian and Jane Clayton, husband-and-wife ecotourists on their way back home to teaching jobs in Essex, England. I had met the Claytons the previous week at La Milpa Field Station, where we shared an evening, guided by Vlad Rodríguez, spotlighting nocturnal birds and mammals in trees and along the trail. This British couple, outfitted with thousands of dollars' worth of gear, might be every conservationist's dream. They spent every holiday traveling the world in order to get firsthand looks at exotic wildlife. Jane said that their next excursion would be to view either giant pandas in the mountains of China or elephants in southern Africa's Gabon rain forest.

As he drove toward Belize City, the sardonic Julian kept up a wry commentary about the roadkill we observed. "There's a fluff of kinkajou," he announced. "That bristly bit is definitely a peccary; and the dodgy muff over there must be a white-nosed coati." His black humor underscored the danger to threatened creatures posed by something as ubiquitous as motor vehicles, a reminder that simply by building roads and driving cars we put wide-ranging animals at risk.

As he drove, keeping an eye out for incautious wildlife, Julian wondered aloud about the proper role of foreign visitors in modern conservation. "Would it be better," he asked rhetorically, "if we stayed in England instead of burning up tons of fossil fuel to get to the ends of the Earth? Perhaps Jane and I could see the remaining wild creatures of the British Isles. We'd certainly leave a much smaller carbon footprint if we simply tramped around the Lake District or the Cornwall coast."

I didn't have a good answer. "We need to do both," I ventured. "It's important to preserve nature close to home, but local people seem to appreciate the value of their wildlife much more when tourists travel great distances and spend lots of money to see it. Without the thousands of foreigners who patronize Cockscomb each year, the sanctuary might've been carved up long ago."

The Claytons knew this, having been to many small countries where preservation competes head-to-head with development. The three of us stared at the Belize countryside, which had given way gradually from limestone crags to the savanna of coastal plain. There was still plenty of room for jaguars here, I thought, but Belize was fast becoming like the rest of Central America, a thin strand of mountainous land crisscrossed by roads and crosshatched with farms. Earlier in my trip I had passed through El Salvador, an overcrowded country where every arable acre was put into production. Jaguars had been removed completely from the landscape decades earlier. Would the same fate eventually befall *Panthera onca* in Belize and Guatemala, for which landless Salvadorans were now fleeing?

Our stomachs launched into a chorus of hungry growls. We had skipped breakfast and were disappointed to find the only two restaurants on our route closed for Good Friday, a major holiday in Latin America.

Unexpected salvation of a sort came at the Belize Zoo, which was open for business and blessed with a snack shop. I rescued Jane and Julian by treating them to a nutritionally questionable meal of candy bars, tortilla chips, and soda pop.

CHAPTER SEVENTEEN

"These Animals Could Become Wonderful Teachers"

THE CLAYTONS HAD NEVER BEEN to this zoo, but I had visited several times and was acquainted with its staff. I had spent a guided afternoon and evening on the premises only ten days earlier. My companions perked up and asked me to show them around.

"This is not your grandmother's zoo," I advised them. Jane and Julian shot me quizzical looks. "You'll see."

Located twenty minutes east of Belmopan, the country's sleepy capital, the twenty-nine-acre Belize Zoo is truly one of a kind. Operated neither by government agency nor as a commercial enterprise, it is an educational venture overseen by the fungi expert, former Iowa housewife, and erstwhile lion tamer Sharon Matola. With some eighty thousand visitors a year, it is the country's biggest single tourist attraction. The eclectic "Zoo Lady," as she is widely known, sets the tone for what some describe as a Dr. Doolittle kind of place. But the zoo's founder does a lot more than talk and sing to the animals.

"Sharon is more at home in the jungle than most of us are in our mother's kitchen," wrote Bruce Barcott in *The Last Flight of the Scarlet Macaw*, a 2008 book detailing Matola's unsuccessful 1999–2005 campaign to block construction of a 150-foot-high dam on Belize's Macal River, which flooded nesting sites of one of the region's most endangered birds. "An American by birth, she's spent the past quarter century in the raw tropical landscape of Central America. . . . She is a strange and enchanting woman."

At the time of the zoo's founding in 1983, the twenty-eight-year-old expatriate was looking after twenty-one exotic, semitame animals that had been featured in a nature documentary made by her friend, neighbor, and one-time employer, the English cinematographer Richard Foster. After Foster's movie project ended, the Baltimore native found herself out of a job. The same fate befell, among others, a jaguar, an anteater, two parrots, and a Baird's tapir—the endangered cousin of horses and rhinos that Belizeans call "mountain cow." Unwilling to release the film's stars into the wild, where she was convinced they would be unable to care for themselves, Matola promised her wards a reprieve.

"I knew these animals could become wonderful teachers," she told me during my first visit to Belize, in 1986. "If I turned them out, they'd either be eaten or starve."

Matola's résumé reflects her background of creative and pragmatic problem solving. She became fluent in Russian at college, ditched her midwestern dentist husband to join a traveling circus as a dancer and lion tamer, sneaked across the Rio Grande with a smuggled monkey balanced on her head, underwent jungle survival training in Panama during a hitch in the U.S. Air Force, and built a following as one of the most popular rock-and-roll disc jockeys in Belize. Matola is also a recognized scientific authority on area wildlife and the author of a series of nature-oriented books for schoolchildren, whom she adores introducing to the outdoors.

"Most Belizeans live in urban areas and have little or no contact with their country's amazing wildlife," she said, by way of explaining her long-ago decision to stick a hand-lettered sign in the ground reading: BELIZE ZOO.

At first Matola raised chickens to feed her captive animals and keep the enterprise going. She zipped around the country on a motorcycle, selling chicken to restaurants as a way of raising operating funds. The Belize government endorsed the zoo in concept but could not support it financially. Since the beginning, key officials have liked Matola's irreverent, fun-loving approach to environmental education, and by 2009 the zoo was bringing six hundred teachers and seventeen thousand schoolchildren a year to its kid-friendly facility. Many students visit in spring for Matola's "April the Tapir" birthday party. The zoo's docile mascot—now a national symbol of conservation—is awarded a cake made of horse feed, grated carrots, and hibiscus blossoms. The humane treatment of April is no exception. Matola's operation prides itself on sensitive care of every animal.

"Most of them are injured or orphaned or a pet that someone grew tired of," she pointed out. "They have nowhere else to go."

During this visit, I was pleased to find that the zoo had several adult jaguars on display in large outdoor pens, including one I had not seen before. An old favorite was C.T., named after his hometown of Crooked Tree, an isolated village and world-renown birders' destination, surrounded by wetlands. He was rescued after his mother, a cattle killer, was shot. "No one knew there was a cub in the picture when C.T.'s mom died," Matola told me, "so we adopted him." Delegated by circumstance to a life in captivity, the handsome neutered male loved to doze in the highest branches of an oak tree growing at the center of his leafy compound.

Sharing C.T.'s space was Ellen, a black-phase female captive-bred at the Ellen Trout Zoo in Lufkin, Texas, and later donated to the zoo. In the right light, Ellen's pelage clearly revealed the rosettes and spots outlined but otherwise hidden in her intensely dark fur. Looking at her was like staring at the optical illusions that delighted me as a kid: one moment Ellen appeared totally dark, the next her coat was dappled with circles and dots. In the adjacent enclosure was newcomer Frankie, a laid-back male who delighted camera-wielding visitors by lounging in a part of his commodious space that resembled deep jungle. Nearby, in their own compounds, were living examples of Belize's four other indigenous felid species.

On the day of my previous visit, shortly before my tour with the Claytons, Sharon Matola greeted me dressed informally in knee-high waterproof boots, khaki shirt, and camouflage pants. Now in her early fifties, she had become one of the country's most widely recognized (and controversial) citizens. Tall and slender, with chiseled features and dark, gray-streaked hair, Matola has a manner that is direct, informative, and sincere. As a passionate conservation activist—whose outspoken style rubs some Belize politicians and business leaders the wrong way—she is respected by many

Figure 17.1. This captive adult male jaguar, named C.T., relaxes in an oak tree at the Belize Zoo. The animal was orphaned as a cub when his mother was shot near the village of Crooked Tree. Photo courtesy of Belize Zoo

for her strong values, bold initiatives and creative flair. Examples of the latter include the whimsical, hand-painted signs Matola posts before each enclosure. "LOOK AT ME," proclaimed a notice before a parrot's cage. "MY SQUAWK IS LOUD! THINK OF THAT BEFORE GETTING A BIRD LIKE ME AS A PET."

As we stopped to hand-feed several animals, Matola gave me a rundown on the facility's resident jaguars. Besides the three animals on exhibit, three wild-caught jaguars were undergoing rehabilitation in a secluded section, away from visitors. Also on site were the aged Pete (retired from exhibition) and three-legged Angel (a captive-born cat reared by Sharon after rejection by the cub's mother). The black jaguar, Ellen, rolled over in a dust bath as I approached with Matola, who clucked her tongue and applauded.

C.T. continued to sleep in his tree, while photogenic Frankie emerged from beneath the shade he favored on the east side of his pen. Ellen and Frankie accepted Matola's raw meat scraps as she handed them through a mesh barrier. A sign advised: "SEE MY SPECIES IN THE WILD AT THE COCKSCOMB BASIN WILDLIFE SANCTUARY."

Matola has fun grabbing attention, but her underlying message is dead serious. "We're barking up the wrong tree as far as conservation goes," she said, as we strolled about the zoo, interrupted by one child after another eager to meet the local equivalent of Willy Wonka. "People in countries like Belize want to make a living. Let's make that possible for them *through* conservation, not *in spite* of it. The Cockscomb, for example, draws tourists who also bring much-needed income to nearby residents. Even if sanctuary visitors don't actually see a jaguar, they're excited simply to know the cats are there."

We entered the animal commissary, behind the zoo's public area. "You'll notice that I call each jaguar by name and give it visual cues," Matola said, refilling a red plastic food bucket with chicken meat. "Big cats like their routines; knowing what to expect helps them feel safe and secure."

Next stop was the "problem jaguar" rehabilitation area, so named because it housed cats that had killed livestock or other domestic animals. Under a program launched in 2003, those suffering losses due to jaguars were asked to call Matola's office and arrange for a "suspect" cat to be captured unharmed rather than shot or poisoned, as is usually the case otherwise. At the time the program started, more than one jaguar a week was reported killed in Belize, though the actual number may have been up to three times higher.

"Predation upon livestock and domestic animals is a major problem," Matola allowed, "and there are few to none inputs to address it. Ours is unique and it is working. I'm so proud that we have three [rehabilitated] cats going to other zoos. They're set to become icons for their species while contributing valuable genetic material to the captive jaguar population overall. How can we look to the future for possible restoration of this species unless a healthy genetic stock exists?"

Matola emphasized that her facility does not breed jaguars, though occasionally a wild pregnant female or young cub arrives. The need for fresh DNA is real, she added, since many captive jaguars in U.S. zoos have unknown ancestry. Without fresh genetic input, such a population runs the risk of producing offspring with birth defects or debilitating health problems. Some cannot reproduce at all.

My escort noted the success of the region's harpy eagle reintroduction program, which the Belize Zoo facilitates. Through the efforts of Matola and such scientists as Ryan Phillips, these charismatic eagles are captive-bred for release in Belize, Guatemala, and Mexico. "It is a success story unmatched," Matola enthused. A similar restoration program, she said, might one day bring jaguars back to places where they, too, have been extirpated.

"The entire country [of Belize] now knows the harpy eagle and wants the bird protected," Matola continued. "But we need a healthy population of a species in captivity in order for such programs to have a chance." Similar approaches involving whooping cranes, Mexican wolves, California condors, and black-footed ferrets have met with varying success in the United States and Canada.

Shouting his name and stopping just outside a protective chain-link fence, Matola rousted Wild Boy from the den he shared with his girlfriend, Zabby, a shy but effective sheep killer. Shot, trapped, and beaten by humans, Zabby does not trust them. Her fine-looking suitor sauntered over and sat immediately opposite the mesh from where I squatted with my camera. My host fed bits of chicken to the jaguar and encouraged Wild Boy to perform the tricks he had learned during six months of captivity. These included rolling over on command and doing "high-fives" against the chain-link with his broad forepaw.

"Jaguars are very intelligent," Matola said. "They like to learn. In fact, they have the largest brain-to-body mass ratio of all great cats."

In response to Matola's prompting, Wild Boy displayed the broken teeth that, sadly, had turned him into a cow killer. I delighted in hearing his many vocalizations: grunts, growls, coughs, chuffs, and throat clearings, all uttered mere inches away. Here was a cat separated from his jungle freedom by only a half-year. The encounter might be the closest I would come to observing a truly wild jaguar.

Matola believes that "a decent respect for wildlife" in the future devolves from teaching youngsters today. To that end, the director has produced hundreds of radio programs, given countless talks, and written three books for children. "My next story will be about Wild Boy," she said, "told from his point of view." Matola launched immediately into this plot summary: "I got into a fight with another male jaguar over territory that we both wanted for hunting. I lost out and my teeth were injured terribly. I had no choice but to hunt animals that were easier to kill, like calves. The second time I did this, a rancher shot me. I wasn't badly injured, and some people came to capture me in a box trap. I couldn't escape, and the next thing I knew I was in a pen at the zoo. Miss Sharon was sitting outside the bars, playing her guitar and singing a song about me."

We moved to an adjacent enclosure containing another adult male. Matola explained that this jaguar, captured after eating one too many dogs, was named after the most popular pet food brand in Belize. Fieldmaster had also lost teeth, which probably had turned him into a dog-killer, and he could not eat large chunks of meat. This elegant creature was even more of a show-off than Wild Boy, rolling on the ground like a playful kitten. He was already scheduled to begin a new life in a Pennsylvania zoo.

"When Fieldmaster came here," Matola said, "there were forty botflies under his skin. They could have killed him." Botflies are a scourge to many mammals in Central America, including humans. The fly injects its larvae beneath an animal's skin, where they grow for weeks before eating their way out. These pests and various parasites can turn an otherwise healthy jaguar into a deathly sick one.

We walked to the sturdy cell-like enclosures where wild jaguars are housed when they first arrive. "We call them 'dens,'" Matola said, adding,

Figure 17.2. The Belize Zoo's founding director, the biologist Sharon Matola, rewards a captive "problem jaguar" with chicken scraps when the dog killer gives Matola a "high five" against the chain-link enclosure. Such jaguars generally are placed on exhibit at the facility or shipped to other zoos. Photo © Richard Mahler

"jaguars love denning areas." This correlates in the wild with their propensity for finding (and using) caves, ledges, and outcrops. The zoo's human-made dens were small yet comfortably furnished with scratching posts for claw sharpening as well as shelves for sleeping and perching. The location was shaded and away from other animals. When a new cat arrived, Matola would sit nearby for hours on end, playing her guitar and softly singing custom-written songs. She hand-delivered food and water, speaking to the jaguar in a sort of "motherese" calculated to lower its anxiety.

"My goal is to calm them down and get them used to people," she said. "This treatment allows the cats to adapt before we release them to the larger pen, where they have enough trees and shrubbery for natural shelter and privacy." The acclimation also adjusts them to other jaguars, a novelty for these otherwise solitary creatures. The process was something a Romanian lion tamer had taught Matola thirty years earlier in Florida, insisting that big cats "need to receive positive reinforcement and learn what's expected of them."

I noticed chord tablature and lyrics taped to the side of the intake unit. It was Fieldmaster's very own song:

I killed them big and I killed them small,
I ate them short and I ate them tall.
One day I wandered into a trap;
Ranchers jumped for joy, you can bet on that.
Jaguar rehab, that's my deal.
This chick can feed me every meal.
She sings to me and she brings me bones.
The hell with cows, I've found a home.

Matola has said her most pressing concern was securing enough room for rescued jaguars. At the time, a female with its cub was about to be trapped near Altun Ha, a Maya ruin popular with tourists. A third jaguar, accused of killing cattle, also awaited capture. Matola was confident that, as always, the needs of every inhabitant would be met. Her operation enjoyed worldwide acclaim and past contributors had included such celebrities as Harrison Ford, Calista Flockhart, Jimmy Buffett, and the late Steve Irwin. British royalty, including Prince Andrew and Princess Anne, had dropped by to pledge support. Much of the funding, however, dribbled in a few dollars at a time from visitors foreign and domestic.

Next stop was the wire-mesh enclosure next to Matola's rickety bungalow office, perched on stilts, where three-legged Angel was waiting. Matola explained that the cat's foreleg had been amputated when it became infected following an altercation with another jaguar. The director extended a hand inside the fencing to rub the feline's soft face and fondle its extended paws. Angel rolled over and stretched, displaying a tummy that begged to be scratched. I could swear that the animal was purring. She appeared to be as affectionate as a housecat, but I knew better than to approach a 180-pound jaguar uninvited. "Don't pet her," Matola advised, "unless you're ready to part with your hand."

❧❧

We walked back for a sit-down interview in Matola's office, just above the reptile house, where a hand-lettered sign pleaded with visitors not to kill snakes on sight, as was the local custom. (Only a small percentage of the country's snakes pose any danger.) Settling into chairs in a sun-flooded room, I noted that the Belize Zoo was now a must-see destination for tens of thousands of people each year. Matola agreed that the facility had become, beyond her expectations, a national institution that alerted both locals and tourists to the importance of protecting nature's hot spots.

"There is no other place in the world I would rather be than smack right here," she said. "This is exciting work and a chance to make a real difference. I love it."

Matola has a life away from the zoo, of course. The director hosts a popular radio show featuring her favorite rock-and-roll tunes and is admired for her fashionable collection of sequined jumpsuits. When she can get away, the biologist leads scientific expeditions into the wildest parts of Belize's interior, where much is still being learned about native wildlife.

"We are very proud of the influence the zoo puts forward," the director concluded. "The purpose of the jaguars at the Belize Zoo is to show visitors the awesome beauty of this animal, yet we also draw attention to the importance of keeping jaguars roaming free and wild. The ecosystems such cats inhabit provide us with so much. Protect animals like these and you protect everything else in an ecosystem."

❧❧

One rarely gets a chance to see a nocturnal animal close up in its preferred dark of night. So when Sharon Matola offered an after-hours look at the Belize Zoo's cats, I accepted immediately. Although wild jaguars are sometimes active during daytime, depending on their personalities and circumstances, they are most wakeful in darkness or at dawn and dusk.

On this warm, oppressively humid evening, I felt an electric current of animal energy inside the zoo. The fruit-loving kinkajous, for instance, were wide-eyed and bushy-tailed. During my afternoon visit, these small, fuzzy mammals had been bundled in a snoozing heap. Reptiles and amphibians slithered visibly in a pond that had appeared lifeless at midday. Bats flitted about, squeaking and diving for insects among the overhanging branches.

The cats were even more impressive. Though two of the "display jaguars" had been attentive during my earlier visit, in the inky darkness they somehow seemed more alien and stealthy. The soft shadows and fusty air only enhanced the cats' inherent mystery. In a phrase, they were more exciting.

"That's C.T.," said the finger-pointing guide escorting me and four other visitors past the sturdy enclosures where Belize Zoo jaguars spend the night. "Over there is Ellen, who shares her compound with C.T. during the day." In the third unit was Frankie, looking more energetic than he had while immersed in his noontime nap.

The three jaguars enter the sturdy cages—Matola refers to them as comfort zones—when the zoo closes each evening. The moves are made for practical reasons. The denlike enclosures give staff a chance to observe, medicate, or dispense vitamin supplements as needed. Because the facility is in a hurricane-prone region, confinements must be safe and secure. No one would want agitated jaguars on the loose during the horrific tropical storms that rake Belize every few years. Installation of the units followed a close call dealt by Hurricane Mitch in 1998, which lashed the country with strong winds and drenching rains.

"During night tours," Matola told me later, "the jaguars are given bits of raw chicken as treats."

But C.T. had not yet been given his meat, and while we watched he threw the jaguar equivalent of a temper tantrum. He snarled and growled, moaned and groaned. At one point, the two hundred–pound cat hurled his entire body against the bars on our end of his "comfort zone," shaking the structure like a small earthquake. I knew that this animal was healthy and well cared for, but as if he were a tearful three-year-old toddler "acting out," C.T. made clear his displeasure at not getting his chicken bits at precisely the moment he wanted them. His performance provided a convincing demonstration of a jaguar's oral range and physical strength. Certainly, every animal in the zoo heard C.T.'s thunderous roar. (Matola later assured me that before our group departed, each jaguar received its bedtime snack and C.T. fully regained his composure.)

"How exciting," exclaimed one of my Canadian companions, a wide-eyed retired schoolteacher accompanied by his wife, a chatty social worker. "We've never seen anything remotely like this in Vancouver." Nor had I, anywhere.

As we moved about, the zoo's other felids offered curious stares and intriguing sounds. We received warning hisses from the margays and jaguarundis, nervous purrs from an ocelot and mountain lion. Another of the lions surprised us with an unexpectedly meek meow, a third with an annoyed growl.

🦅🦅

Having spent considerable time, effort, and money in search of a wild jaguar, I felt reasonably content after my day and night tours of the Belize Zoo. I savored vindication of my belief that these cats were, indeed, among the most charismatic creatures on Earth. I had now been close enough to them to see their ornate coats, hear their commanding voices, and bear witness to their prowess. This wasn't the same thing as encountering a jaguar roaming free, but it still raised goose bumps.

Zoos are imperfect institutions, but they are still the only place the average person is likely to see a jaguar. In 2009, according to the American Zoo Association, about 130 jaguars were held by facilities across the United States and Canada. Approximately 200 others were exhibited in other countries by various zoos and parks. Some 330 "display" jaguars may seem like a large figure, but *Panthera onca* is far outnumbered by captive lions and tigers. About 450 tigers and an even greater number of lions are on exhibit in the United States and Canada alone. Hundreds of others are shown by facilities in other parts of the world. (The American Zoo Association's "studbook keeper" closely monitors how many exotic cats are kept in captivity, where they are maintained, and their lineages.)

Life is different for a zoo jaguar. For instance, it may easily survive into its mid-twenties, about twice the felid's expected life span in the wild. The cat may produce many more offspring, too. Adult females in captivity may give birth from about three years almost into old age.

Another difference is that a high percentage of captive jaguars have dark pelage rather than the black-spots-and-rosettes-on-buff pattern typical of an estimated 94 percent of their counterparts in the wild. Why? "Zoos care to exhibit the unusual or spectacular in order to bring attention to certain species," Roberto Águilar, the veterinarian who directs conservation and science for the Phoenix Zoo in Arizona, explained. Melanism—the genetic condition that darkens background fur—"is relatively common in exhibit animals" because it is "of general interest to the public, who wish to see a 'different' animal."

Jaguars can be bred relatively easily in captivity, Águilar added, "ergo the disproportionately large number of jaguars kept under poor-to-terrible conditions in Latin America, not in zoos [but] usually in [captivity related to] illegal animal trade."

Keeping such a cat is a daunting undertaking. Jaguars are large, active animals that demand lots of room, plenty of raw meat, and a high level of commitment to ensure long-term care. The cats are intelligent and trainable, though not particularly trustworthy in a circus ring. They can be ornery—and are always stronger and faster than their human handlers. "If a [captive] jaguar decides to turn on you suddenly, there's really not much you can do," said Barbara Diceley, who with her husband, Rob, runs Leopards Etc., a California facility that raises exotic cats for live presentation in "educational adventure" programs. "We won't keep jaguars. Yes, you can handle them when they're cubs, but what happens after that? They spend the next twenty years sitting in a cage."

Through the Jaguar Species Survival Plan of the Association of Zoos and Aquariums (AZA), thirty-nine zoos have dedicated themselves to long-

Figure 17.3. Often (and inaccurately) referred to as black panthers, about 6 percent of wild jaguars, mainly in South America, have genetic melanism. Such jaguars may appear black from a distance, but faint rosette and spot patterns can be seen in the coat upon close inspection. Photo © Carol Farneti Foster

term management and conservation of jaguars. Participating zoos help maintain a genetically healthy population and make visitors aware of the key role this particular felid plays in habitats of the Americas. But well-meaning zoos and wildlife parks can do only so much. They are limited by funding, space, and other constraints. Moreover, captive jaguars are not able to present—much less retain and pass on—the full range of behaviors and knowledge of wild animals, particularly if the cats are a generation or more removed from their natural environment. Some "exhibit jaguars" come from long lines of captive animals and may be separated from the wild by five or six generations. A cub spends two or three years learning from its mother how to fully *be* a jaguar, and much of what she teaches her offspring remains unknown to humans. With the premature death of each wild jaguar, a certain amount of unique, location-specific information is lost forever.

Richard Foster speaks knowledgeably about essential differences between wild and captive jaguars. A native of England and a longtime Central America resident, he is the cinematographer whose nature film yielded the group of captive animals around which Sharon Matola built the Belize Zoo.

Richard and his U.S.-born wife, Carol Farneti Foster, have lived—quite literally—among jaguars for decades. Since the early 1980s their home has been directly adjacent to the zoo, and it was initially on their property. The jaguars padding about forested enclosures in their backyard starred in eight nature films over a twenty-five-year period. Carol, a wildlife biologist, is a full partner with Richard in these ventures in such roles as producer, director, researcher, and photographer. The savanna surrounding the couple's home is thick with natural grasses, dotted with pines and occasional patches of tropical forest. This landscape harbors wild jaguars, though most congregate several miles away in an isolated swamp. "We hear or see signs of the cats once in a while," Richard said, during my brief tour of the compound. The filmmaker added that Alan Rabinowitz had come here as a houseguest before his Cockscomb sojourn. The Fosters became his good friends as well as mentors.

"We helped Alan 'catch' his first jaguar," recalled Carol. "Unfortunately, the cat was dead. Someone ran into it [on Belize's Western Highway], and the three of us went out to examine the body." The "someone" was a high-ranking government official who infuriated Rabinowitz by trying to abscond—illegally—with the jaguar in order to keep or sell its valuable pelt.

The ground floor of the Fosters' modest, airy home was divided into "his and her" studios. Framed oversize nature photos adorned walls above Carol's workspace, while Richard's zone was cluttered with reels of film and other paraphernalia of moviemaking. The couple led me to an upstairs dining table of fine tropical wood, at the center of which an orchid bloomed extravagantly in lavender and scarlet. We sipped fresh chilled lime juice in the furnace heat of late afternoon. I asked what, in their respective opinions, jaguars were *really* like.

"It's hard to generalize because each cat has a unique personality," said Richard, an easygoing man with a kind face, stocky physique, and shaggy, receding brown hair. "They definitely are creatures of habit, but they're also very curious." He said that jaguars prefer to use existing trails and logging roads when available, otherwise taking advantage of pathways of any kind. Yet they seem to deliberately travel in the wee hours, when people aren't around. "Above all," he said, "the jaguar is a loner, a free spirit, and extremely observant."

Carol, an animated brunette with tight curls and constantly gesturing hands, said that she particularly admired the cats because "they are very smart and have great memories. They recall things so completely that they notice when even the smallest item is out of place within their domain." The Pennsylvania native observed that the cats "can tell at a glance whether they're going to like you. . . . They remember behaviors, too, especially perceived insults. If you shoot a jaguar with a tranquilizer dart, it'll neither forgive nor forget. Whenever you come around after that, it'll growl or disappear."

The couple conceded that they have experienced only limited success filming jaguar footage in the wild. Richard had seen only three jaguars "in situ" during his lifetime: two in Belize and one in Venezuela. "Trying to record them that way is not cost-effective," he said. "One might spend weeks trying to get a jaguar on film and be lucky to have only thirty seconds in the end—of a cat running away." As is the case with many nature films, most of his and Carol's productions have starred captive cats.

Getting such an animal to "act wild" is challenging. Richard outlined the difficulty in getting captive jaguars to behave "naturally" on camera. For a National Geographic documentary he nearly had to get on all fours to demonstrate explicitly to a captive cat how to catch a fish. Not used to hunting or fishing, the animal came through only after lengthy rehearsal. Richard lamented the fact that, due to their unpredictability, wild jaguars have rarely been filmed engaged in some activities easily observed among other species, including stalking as adults and teaching cubs as parents.

"Yet in many ways it seems like jaguars are simply overgrown house cats," Richard continued. "They wash themselves the same way. They love to lie in the sun when the weather is cool."

The filmmaker was clearly mesmerized by jaguars. "Why?" he asked himself aloud. "Because they live at the epicenter—the very heart—of the forest. And another reason is their spotted coats, which are stunningly magnificent." Richard stood up and wandered to the kitchen in order to refill his juice glass.

Carol leaned forward in her seat as her face congealed into an earnest expression. "When I see a jaguar, I feel sense of awe at their power and beauty," she said. "But I have no fear in being near a wild one in the forest." According to Carol, jaguars are naturally inquisitive, like other cats, but seem to prefer observing humans from a safe distance. They will not reveal themselves in the wild unless forced to do so. Fortunately, she and Richard could make do with willing subjects that had been among humans most, if not all, of their lives.

The Fosters' parade of backyard jaguars had included current zoo inhabitants Pete and Angel, along with past resident Mick Jaguar. In residence at the time of my visit was eight-year-old Boo, a family "pet" as well as film star.

Richard returned and beckoned me to follow him downstairs and then outside. He walked me away from the house for an introduction to Boo, who lives in a patch of native forest surrounded by a high chain-link fence the Fosters had built specifically for carrying out jaguar-themed photography. Strategically placed holes in the enclosure provided unobstructed access for camera lenses, yielding depictions of the cats in naturalistic settings. Boo was kept in a cozy management den at night and during the day enjoyed a mesh-enclosed "run" that allowed him to sprint through the trees in a fabricated tunnel. I had seen Boo perform in several Foster productions, and it was a treat to admire him from only a few feet away. The big male made threatening snarls as he rushed back and forth overhead, showing agitated curiosity at my unexpected arrival. Richard calmed the cat with reassuring words.

"I mean no harm," I told Boo in a low voice. "I just want to have a look."

"That's right," Foster cooed. "He won't hurt you. And you already know I'm a pushover."

I stared at Boo with undisguised pleasure while a pair of dogs barked playfully below the elevated jaguar run. The dogs seemed to pursue a familiar game with the cat, which snarled and hissed in mock anger while the canines raced around below the aerial cage. Boo's attention shifted completely from me to the hounds while Foster slipped into a philosophical mood. Without prompting, he confided his distress at the continuing loss of jaguar habitat. "Humans are so narrow-minded and short-sighted," he complained softly. "Meaningful wildlife conservation will occur only if local people can see it providing them with economic benefits, or if rich [benefactors] or nonprofit groups buy large pieces of land and keep paying to protect them."

The expat Brit gazed solemnly at his lovely captive. "Otherwise . . . ," Foster began. The sentence hung incomplete in the air. We watched the jaguar at play for several minutes as the jungle night descended quickly around us. I would have been happy to sit on the ground for hours, observing the cat in its hypervigilant evening mode. But it was time to go.

CHAPTER EIGHTEEN

"It's Good if It's Dead"

COSTA RICA IS A PEACE-LOVING Central American country famous, justifiably, for its scenic beauty and diverse ecology. Nature-oriented tourism pumps hundreds of millions of dollars each year into the nation's economy. But similar conditions exist here as elsewhere in the Western Hemisphere's subtropic and tropics. The population is rising, agriculture and urban centers are expanding, and pressures on wildlife wrought by hunters and exotic animal traffickers are increasing.

"Good luck finding a jaguar in Costa Rica," one Central American scientist told me, shortly before I flew to the sprawling capital of San José. "You'll need it. The tourism brochures are deceptive."

Minutes after my arrival, before I even exited the airport, I began to understand. Here, as in virtually every town I visited south of central Mexico, the jaguar's image was ubiquitous. The elegant feline had become a clichéd icon promoting ecolodges, tour companies, restaurants, souvenir shops, art galleries, and a host of products and services. Its whiskered face seem to stare at me balefully in every information booth, hotel lobby, and travel agency. This symbol of Costa Rican wilderness confronted me many times each day. But where was the cat?

Three of the country's largest national parks—Tortuguero, La Amistad, and Braulio Carrillo—were "suffering serious losses of their jaguars," according to Eduardo Carrillo, a biologist and regional authority on *Panthera onca,* whom I sought out at Costa Rica's national university in Heredia. In six years "jaguar numbers in Corcovado [National Park] dropped from about 150 animals to a mere 30 or 40." Farmers and ranchers were killing the cats at an alarming rate, and massive hunting of prey animals was fast eliminating the jaguars' natural food supply.

Carrillo and other experts conceded that jaguar populations had been decimated since 2000; the cats, once abundant, were now seldom seen. The species was still known to occur in isolated parts of government-protected reserves, including the Santa Rosa and Río Macho national parks, as well as lower levels of the Cordillera Talamanca. The consensus among those I spoke with seemed to be that the most likely place for a casual encounter was Corcovado, located in Costa Rica's far south.

Corcovado is a gem of nearly pristine rain forest covering much of the Osa Peninsula, a hilly thumb of land thrust like a jade-green island into the Pacific Ocean across a saltwater gulf from Panama. Described by *National Geographic* as "one of the most biologically intense places on Earth," this hard-to-access park protects at least 375 varieties of bird, 140 types of mammal, and 116 amphibian and reptile species. Its broad expanse of humid tropical jungle is home to many animals that have become endangered or extirpated elsewhere, including various parrots, tapirs, anteaters, sloths, and monkeys.

The Arizona biologist Emil McCain, who spent seven months on the peninsula studying jaguars—at times coming within a few feet of them—recommended the Osa to me as a great venue for observation of tropical fauna. He noted that scientific research stations within Corcovado welcomed small numbers of visitors and the handful of commercial lodges just beyond the park's borders specialized in wildlife viewing. Local natural history guides, he assured me, were excellent. What's more, McCain's mentor, Eduardo Carrillo, had conducted a series of earlier jaguar studies in Corcovado and had seen the cats regularly as they trotted along Pacific Coast beaches.

The bespectacled Carrillo, a charming University of Massachusetts graduate who seems to wear a bemused expression at all times, also observed jaguars ambling along the leaf-littered jungle floor, lolling atop cliffs, swimming across rivers, and leaning casually against the buttressed roots of a towering ficus tree. But it was the jaguars' pursuit of giant sea turtles that most impressed this rosy-cheeked researcher. It was more proof that these felids have some unusual predilections.

Where other great cats seem to regard armor-covered prey animals as not worth the trouble of the killing and the flesh extraction, jaguars thrive on them. Rabinowitz, for example, found armadillos a common part of the cat's diet in Belize. In coastal and riverine Central America, where turtles are abundant, jaguars frequently catch and crack open the hard coverings of these animals by using their strong jaws and sharp canines. The felid may also simply kill the animal, reach inside its shell, and scoop out its flesh. As Louise Emmons speculated in a 1987 article in *Behavioral Ecology and Sociobiology,* such behavior may be adaptation held over from the mass extinctions following the Pleistocene epoch, during which armored reptiles and amphibians became easier to find than the many land mammals that eventually went extinct.

Carrillo determined that jaguars at Corcovado fed mainly on marine turtles, along with white-lipped and collared peccaries. (The collared peccary, locally dubbed javelina, is the same species presumably stalked by desert jaguars of Arizona and New Mexico.) The Costa Rican biologist confessed in a 2007 *Natural History* magazine article that his findings took him aback, "because it is clear that a jaguar can eat any animal that crosses its path." On reflection, Carrillo realized that "opting for peccaries and sea turtles makes sense for jaguars: adults of both kinds of prey are easy targets and weigh between 80 and 90 pounds, so they provide a good deal of energy in one fell swoop." Weighing the number of calories expended versus the number potentially rewarded by a prey animal is something apex predators may do instinctively before an attack.

But both Carrillo and McCain worried about the geographic isolation of the Osa jaguars, surrounded as the cats were on three sides by ocean. Although the animals were thought to represent the largest density of *Panthera onca* left in Costa Rica, the fourth side of their territory was being developed rapidly for farming, logging, ranching, and tourism.

Between 2000 and 2007, Carrillo estimated, the park's peccary population had declined 60 percent, primarily because of illegal hunting by humans encroaching from the reserve's fringes. "The hunters use high-caliber automatic rifles, such as AK-47s," wrote Carrillo, "which can kill as many as 50 animals in a few minutes." During the same six-year study period, estimated jaguar numbers in Corcovado dropped by more than two-thirds. "Better protection of the corridor that connects the park with other protected areas is essential," the scientist concluded, in order "to guarantee long-term survival of the jaguar in Costa Rica."

The demise of jaguars on the Osa Peninsula, Carrillo warned, would probably bring on the secondary loss of other animals, spurring a cascade of interrelated events that would lead to even more changes in the structure and composition of the rain forest. It is a process under way throughout the tropical Americas, as elimination of large carnivores allows some other species to flourish and others to decline or disappear entirely.

"But thankfully, Corcovado still is not easy to get to," understated a resident of Puerto Jiménez, the Osa Peninsula's only town, shortly after my arrival. "That is what protects it—so far."

I endured a grueling ten-hour bus ride from San José and emerged in a place that for the most part lacked electricity, telephones, Internet service, groceries, indoor plumbing, running water, and paved roads. Most visitors were hard-core ecotourists or die-hard surfers. The Osa was Costa Rica's "Wild West," though swiftly adding the amenities of modern civilization.

After a night in Puerto Jiménez, where I awoke to a flock of squawking scarlet macaws just outside my window, it was a short ride to Dos Brazos del Tigre (Two arms of the jaguar). This scruffy community—once the center of local gold-mining—was tucked along bluffs overlooking the confluence of the two main branches of the Río Tigre. The silvery river is named for the once-abundant cats found within its upper watershed, animals seldom seen since a gold rush brought hundreds of fortune seekers to the area during the 1930s. Undeterred, I was advised that patient exploration might yield a glimpse of *Panthera onca,* or at least its tracks.

The rattletrap *combi* (public bus) dropped me in front of the Dos Brazos elementary school, the front wall of which was decorated with a crude, colorful painting of a Bengal tiger. The artist was apparently more familiar with Old World felids than those that once prowled the immediate neighborhood. The back-of-beyond village was nearly deserted in the siesta-hour doldrums, and many houses looked abandoned altogether. Dos Bravos had never recovered from volatile gold prices in the 1940s and 1950s. The establishment of Corcovado National Park placed restrictions on miners and loggers, prompting many residents to move away. A substantial rise in precious metal prices had lured a few back.

I followed the south branch of the tree-shrouded river away from Dos Brazos, wading across its rushing knee-deep highwaters to reach the upstream Bosque del Tigre Lodge. A restaurant manager in Jiménez had recommended the place as a forest-bound place to stay, with a strong nature orientation. The secluded inn was run by Abraham, a Costa Rican and former gold miner, and Liz, his American wife. I met them in front of their hand-built home, where the path ended after a shallow ford. Abraham, a loose-limbed fellow with a ready laugh, told me that he had seen tracks of a jaguar on a nearby trail two weeks earlier. He added, with a grimace, that all large mammals had been in steady decline during the ecolodge's dozen years of operation.

"On top of the illegal poaching and timbering, we recently had eleven years of drought," volunteered Liz, whose skin had the burnished look of bronze sculpture. "That dry spell was followed by way too much rain. The weather's been very weird, and it's been hard for the forest animals to adapt."

The couple was packing hurriedly for a vacation in the United States, so Liz directed me up the road to Cabinas Los Mineros, a onetime brothel converted into a simple restaurant and rustic hotel by Suzanne, a flamboyant strawberry-blonde expatriate of French-Dutch-American heritage. During the heyday of mining, the place had been a rowdy bar. One of the first tidbits Suzanne offered upon my arrival was that two men (on separate occasions) had been murdered in cold blood at Los Mineros. "One *cabrón* was shot on the stool where you're seated," she announced. "The other took a machete in the gut as he walked in the door." Such questionable conduct explained the two jail cells I saw at the end of a row of *cabinas* (cubicles) previously used by resident prostitutes. My residence for the next three nights would be a bedroom once belonging to a prostitute.

Although Suzanne, an inveterate hiker, had never seen a jaguar during her many years in Dos Brazos, inquiries suggested that she was in the minority. One villager showed me a necklace of jaguar teeth he wore to give himself *"el poder del tigre,"* jaguar power. Another described the medicinal use of rendered jaguar fat as a healing poultice for sores and wounds. A third informed me that a jaguar had killed four pigs on a nearby farm in recent weeks. When I asked a fourth man, an Oregonian married to a Costa Rican, how most locals felt about the spotted cats, he raised his arms, squinted an eye, and pantomimed aiming a rifle: "They think, 'It's a good jaguar if it's a dead jaguar.'"

By now I anticipated this provocative spectrum of opinion. It was obvious that prevailing attitudes toward jaguars were embedded deeply in location-specific customs, cultural values, and self-interest. For sentimental folks in temperate latitudes, the creature was often a romanticized phantom or an icon of rain forest preservation. In the United States, many were appalled at the disappearance of such handsome cats. But for people actually dwelling in jaguar country, those who felt their livelihoods and well-being threatened, the jaguar was perceived as a nuisance they would be pleased to lose. Only a small percentage in either region seemed motivated to find practical ways of managing *Panthera onca.* Caught in the middle were the scientists and conservationists I had been meeting, who seemed sympathetic to polarized points of view regarding jaguars but frustrated that the felids they wanted to study were shrinking so quickly in number and incidence.

I set off after lunch toward the southeast boundary of Corcovado, which occupies the western slope of the hilly peninsula. I followed an unpaved road past the last of the village's dilapidated houses, then through a field of organic pineapples, looking like a grid of biblical crowns of thorns. The track narrowed into a vague path before descending through a bramble of nettles and bamboo to the Tigre's north fork. The waterway swirled clear and fast through tall second-growth rain forest, too thick with deadfall

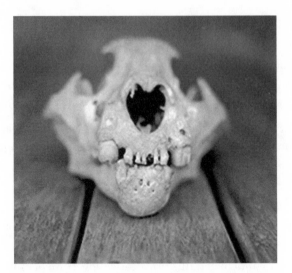

Figure 18.1. The skull of a Central American jaguar killed by a rancher after it preyed upon domestic animals. Note the cat's worn canines and the presence of embedded shotgun pellets. Such factors may have led the jaguar to stalk livestock and dogs rather than its usual wild prey. Photo © Richard Mahler

and sun-hungry shrubs on either bank to allow for bushwhacking. The Río Tigre would be my path onto the low hills that formed the peninsula's backbone. I stood on a sandbar and stared at what lay ahead: an aggressive stream diverted at random by tree trunks, eroded boulders, and polished cobbles. I picked my way through this maze, alternatively wading and climbing, walking and scrambling.

There was no signpost to mark my entry into Corcovado. It did not matter. I had gone beyond the last house, the last road, the last trail. The only indications of human presence were occasional bits of battered, rust-stained mining gear left by gold miners who had panned and sluiced the Tigre for a century. Before modern times, Costa Rica's indigenous tribes had found thousands of nuggets here. These chunks of gold were smelted and molded into such ceremonial objects as the exquisite jaguar-themed ornaments I had seen at San José's Museo de Oro. (I had spent a half-day earlier in the week examining the golden ceramics, jewelry, and stone carvings enshrined in the nation's capital.)

At one point I discovered a makeshift shelter that a miner had only recently abandoned. Ragged lengths of bamboo were strapped together with jungle vines, with a piece of corrugated metal serving as a roof. A rusty tin of tuna and a discarded rum bottle completed the bleak tableau.

I trudged ahead, admiring a gorgeous waterfall cascading from a side canyon, then an ankle-thick vein of rock sparkling with quartz. At each of the many bends in the Tigre, I craned my neck to see whether the river's namesake wandered along the exposed embankment ahead. During hours of searching, I saw no animal larger than a turtle and a heron. With the sun setting and mosquitoes circling, I retraced my steps back toward Suzanne's cold French champagne and hot roasted chicken.

Over the next several days, I followed both branches of the Río Tigre in search of a jaguar. Heeding a suggestion from a Dos Brazos resident, I located a nearly vertical trail leading away from the north fork through the jungle and into the hills. This track eventually brought me well inside the high-canopy forest of Corcovado. As on the river, I encountered plants and animals of astonishing variety, including fragrant orchids and iridescent butterflies, but no member of Costa Rica's six resident felid species. I hiked before dawn, after sunset, and in between, with no success.

Meanwhile, the weather posed formidable challenges. It was the end of the dry season, when midday temperatures soar to well above 90 degrees. Humidity climbed each afternoon to 100 percent as anvil-topped thunderheads built to colossal dimensions. Even at dawn, a hike left me drenched in sweat with eyeglasses fogged and camera slippery. Trekking after lunch meant exposing myself to violent winds and stinging downpours, during which it was difficult to move—and impossible to see wildlife.

❧

After six weeks on the road in four Central American countries, I was ready for a break. Experience has taught me that a few days of rest when traveling provide a healthful boost to the immune system and balance to the psyche. A chance recommendation led me to the perfect refuge: a lovely off-the-grid lodge at a beach on the Golfo Dulce. After taking a bus to Puerto Jiménez, I threw my gear into the back of a local "taxi"—otherwise known as a pickup truck—and headed for Cabo Matapalo, at the southernmost tip of the Osa Peninsula.

I became fast friends with Ojo del Mar's German owners, Nico and Mark, and a warm-hearted Berlin family of four: Sebastian, Nina, Pablo, and Dara. After a delicious dinner served beneath an enormous *palapa*—a thatched-roof, open-sided structure—I was assigned to a beach cabaña and drifted to sleep under a diaphanous mosquito net, serenaded by clicking land crabs, howling monkeys, mango-scavenging coatis, and droning insects.

The next few dawns, days, and dusks were spent exploring the overgrown trails of two nearby nature reserves, where I was assured that jaguars had been seen in recent months. Monkeys and smaller mammals were much in evidence, along with neon-bright frogs and orange-violet crabs, but despite long hours of concentrated searching I encountered neither cats nor cat tracks. As at Dos Brazos, I was soaked daily by monsoon rains that made trails as slick as salamanders and drove most creatures into seclusion by noon. Knowing that jaguars patrolled the local beaches in search of marine animals, I had high hopes that I might encounter a cat consuming a meal of fresh turtle. No such luck.

Evenings at Ojo del Mar passed in lively conversation over family-style meals. Mark described his occasional sightings of an ocelot that lived nearby, and Nico told me of the abundant wildlife she witnessed on regular walks to her part-time job as a masseuse at a neighboring lodge.

Over dinner one night, the Germans and I discussed the double-edged sword of human intervention in nature. We reached a consensus that population growth, technology, and resource use contributed to a global warming phenomenon that was disrupting ecosystems.

"The ramifications of these climate changes are not well understood," said Sebastian, a medical doctor with great interest in ecology. "It is clear that large, niche-adapted predators such as polar bears are being affected negatively as they struggle to adjust."

I pointed out that the impact of global warming on jaguars might be subtler than that experienced by animals near the poles. Some observers had speculated that it is manifesting through shifts in the cat's prey base. They contend the increased sightings of *Panthera onca* north of the U.S.-Mexico border since 1996 may stem directly from the growing incidence of peccaries, coatis, Coues deer, and other mammals associated with ecosystems that are generally warmer and wetter. Such species, eaten routinely by jaguars, are occurring farther north in California, Arizona, New Mexico, and Texas than at any time in human memory. Even wild parrots—once an integral part of the Southwest's ecology—have been seen in New Mexico, while avid birders also flock to southeastern Arizona in order to glimpse eared trogons and flame-colored tanagers, subtropic species found nowhere else in the United States. Some scientists are cautious about taking this argument too far, I noted, since jaguars occurred farther north in centuries past even when the region's climate was cooler and damper.

Based on what we bandied about that night in Costa Rica, climatic changes may portend an ominous impact in Central America. Beginning in 1995, it seems, increased hurricane activity has brought many intense, devastating storms to Mesoamerica and southwestern Mexico. Corcovado had suffered excessive rains the previous autumn, making it impossible for many flowering plants to blossom and bear their fruit or nuts. Mass starvation of many animals followed, including scores of monkeys. Such mammals are an important food source for jaguars and other cats. The torrential rains of late 2005, scientists speculated, were caused by warmer than normal Pacific Ocean temperatures, which influenced ocean currents and wind patterns. Ironically, the monsoons followed an extended period of record-breaking drought that had its own dramatic impact on Costa Rica's flora and fauna.

On another evening our dinner circle was joined by Gabriela, a *Tica* (Costa Rican woman) who relayed tantalizing news about recent jaguar sightings on a nearby road. She recommended a naturalist guide who could take me deep into Corcovado, but when I caught up with the fellow, he confessed that he had seen only mountain lions—not jaguars—in the preceding two months.

Gabriela also suggested I try my luck at La Amistad, a binational park along the contiguous spine of Costa Rica and Panama, where one can gaze down upon both the Atlantic and Pacific after morning mists part. As the largest protected area in the region, this sanctuary was said to have many cats. This was in part because La Amistad is extremely rugged and almost pristine: it boasts almost no accommodations, services, well-marked trails, or road-accessible entry points. Declared a World Heritage site by the United Nations, the so-called Friendship Park was believed to host healthy populations of all felids indigenous to Costa Rica, including the tiger cat (*Felis trigina*) as well as jaguars. But the dry season was over and travel would be difficult, time-consuming, and expensive. I put La Amistad on a list of places to visit during a future trip.

On one of my last evenings in Costa Rica I met a man who lived at the end of a dirt road, where he bred for export a few of the peninsula's 221 species of butterfly, including the stunning electric-blue morpho. An expatriate, John had fashioned a simple life in a place that was simply beautiful. He was content with the Osa's shortage of electricity and prayed that power lines would never extend to his obscure corner of the peninsula. The installation of streetlights would disrupt the habits of butterflies, moths, and other insects, John explained. Television viewing, he worried, might replace the informal dinner parties and cocktail hours that served as a common social currency and evening entertainment. We sat down, sipped the juice from a freshly slit coconut, and spent an hour swapping stories.

"Did you hear about the environment minister's escapade?" the farmer wanted to know. I shook my head, signaling him to continue.

"The guy was on an Easter vacation hike deep inside Corcovado," John explained. "The minister said he was attacked by an angry mother tapir protecting her calf, so he ran away and got lost. [Tapirs are rhinoceros relatives that can weigh up to seven hundred pounds and, though normally docile, will attack in defense of their young.] This fellow spent two nights alone in the park. News updates about the search to find him flashed across the country. Eventually, the guy headed west and hit the Pacific. Smart move. He was plucked from the beach by a rescue helicopter."

"Was he all right?"

"Yeah, just hungry and dehydrated. But the amazing thing was that instead of complaining about his ordeal, the minister said it was one of the best experiences of his entire life. He *loved* it! The San José bureaucrat told reporters he was amazed at the lushness of Corcovado and the diversity of its wildlife."

"Sounds like your minister of the environment will never be the same," I ventured.

"*Y gracias a dío por eso,*" John said with a nod. "And thank God for that."

Eight months later, Costa Rica President Óscar Arias affirmed his commitment to raise $32.5 million to improve protection of Corcovado. The money would supplement $19 million raised privately and through nonprofit groups to help hire sixty-seven additional park guards, acquire an additional three thousand acres, and strengthen community participation in Corcovado's management.

"Our studies show that the peccary population has recuperated," Eduardo Carrillo declared in 2007. "But it will take some years before the jaguar population begins to make a noticeable recovery in Corcovado."

CHAPTER NINETEEN

"A Flagship Species for Conservation"

"JAGUAR HABITAT [THROUGHOUT the animal's historic range] has decreased overall by more than 50 percent in the last century," the pioneering researcher Alan Rabinowitz warned in December 2006. "The broad vision is a jaguar corridor from Mexico to Argentina."

In an interview for the National Geographic Society's podcast, the scientist said that ongoing research suggests that at least some jaguars move unobtrusively through human-disturbed landscapes, such as farms and ranches, en route to the less-inhabited areas where they prefer to mate and hunt. Whereas the paradigm for conservation once was "to find good areas of habitat and lock them up," Rabinowitz continued, "we will not save jaguars [this way] in the long term. We will lose everything that makes them truly wild. [In order] to save them forever we need to save not just postage-stamp core areas but inviolate corridors and larger landscapes that are connected."

Rabinowitz, who in 2008 became head of the Panthera Foundation, a nonprofit group devoted to conservation of big cats worldwide, has proposed that national governments, in cooperation with business interests and using up-to-date information provided by scientists, take steps to protect known corridors and either pay or give tax credits to property owners in return for use of their land by *Panthera onca.* "I think this is going to be the toughest thing I've ever attempted," the scientist conceded, "but the most meaningful in many ways and a model [for conservation] unlike any other."

Convinced that precious little lasting protection had been secured for the species since he first studied it, the man the *New York Times* once dubbed "the Indiana Jones of wildlife science" is adamant about the need for well-established, government-sanctioned travel corridors between "islands" of protected habitat known to harbor jaguars and other large predators. He expressed the fear that without the capacity to remain genetically diverse, jaguars in isolated pockets would become too inbred to flourish. Without the genetic integrity provided by fresh DNA, pregnancies are more difficult and both birth defects and disease more likely. Such a fate already threatens other shrinking large-cat populations, notably the critically endangered Florida panther, which had dwindled to fewer than thirty individuals in the 1980s before beginning a slow, arduous comeback. Fewer than one hundred were said to survive in 2009. Similarly, the number of Asiatic cheetah may have dropped so low that healthy reproduction is in jeopardy. In 2009 fewer than one hundred Asiatic cheetahs were said to remain. Groups this small are particularly vulnerable to poaching, prey loss, traffic deaths, and such catastrophic events as droughts, hurricanes, and wildfires.

Rabinowitz's convictions about the need for aggressive habitat protection and assured connectivity, now widely accepted in the research community, are based on the premise that without access between and among scattered breeding populations, such a species may be at long-term risk of dying out. Scientific studies suggest genetic exchange from a single new jaguar among fifty others will, over a century, preserve a breeding population's integrity much more effectively than will occur in an isolated group of one hundred jaguars. In general, large felids need plenty of room in part because it takes lots of protein to sustain them. As the naturalist George Schaller has pointed out, jaguars seem seldom to occur in high densities, even in ideal situations.

Without further study and data, such theories remain untested. A debate related to the corridor proposal concerns whether *Panthera onca* breaks down, as many widely dispersed mammals do, into subspecies. As taxonomic rankings, subspecies may vary from one another in genetic terms. Sometimes they occur in geographic clusters separated from others of their kind by distance or isolated by such physical barriers as mountain ranges, water, or deserts. But subspecies also may differ from other populations in an observable way, such as size and coloring, yet not be isolated from them reproductively. As one observer put it, "a subspecies is evolution in action," an animal "halfway to becoming a new species."

Some biologists believe that jaguars, which tend to disperse widely when afforded the chance, have been exchanging their genetic material over wide geographic areas for millennia, allowing them to remain essentially the same species from Arizona to Argentina. According to the ecologist Peter Warshall, however, some regional characteristics can be noted. For example, the black marks adorning the flanks of South and Central American jaguars tend toward rosettes alone (the "pintada," or painted, pattern), while northern cats display more dots with their rosettes (the "mariposa," or butterfly, variation). Another gene-based characteristic, as noted previously, is the melanism that occurs primarily among South American jaguars.

In 1990 the Wildlife Conservation Society originated a proposal to create and maintain a big-cat corridor snaking without interruption from Canada south to Patagonia. Initially oriented toward mountain lions and called Paseo Pantera, the project was relaunched in 1997 as the Mesoamerica Biological Corridor, with backing from each of the region's governments as well as the European Union, World Bank, and conservation groups. In May 2006 environment ministers representing Mexico and the seven nations of Central America met in Panama City for the Second Mesoamerica Protected Area Congress and agreed formally to establish such an integrated system of protected areas and wildlife corridors.

Physical movement corridors have become important tools in many aspects of wildlife management. A few animals—notably bear, deer, elk, pronghorn, and caribou—are particularly tied to migration patterns easily disrupted by roads, pastures, towns, and logging. Ski resorts, spas, artificial waterways, and golf courses also sometimes block passage.

Creative solutions can be effective, however. Special overpasses in Canada's Banff and Jasper national parks allow animals to travel safely from one side of the Trans-Canada Highway to the other. Mesh fences

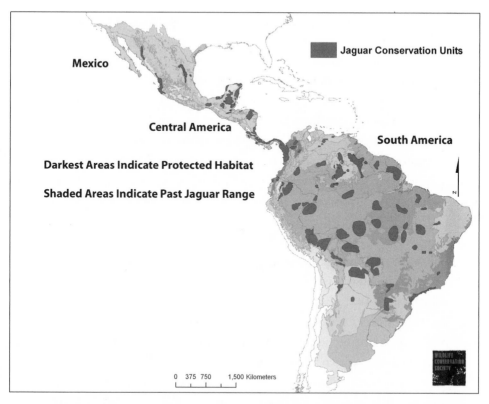

Figure 19.1. The "conservation units" on this map indicate areas of known jaguar habitat that scientists seek to link through protected travel corridors aimed at ensuring the cat's genetic strength and diversity. Illustration by Kathy Marieb, © Wildlife Conservation Society

and other artificial barriers are used to keep animals off this busy highway and nearby railroad tracks. These also funnel creatures to the crossovers, further reducing roadkill. The overpasses resemble sloping hills covered with grasses, shrubs, small trees, and wildflowers. Earthen berms hide the highway and railroad from the animals, who otherwise might be reluctant to use them. Scientists monitor activity on the overpasses with motion-sensitive cameras and "track pads" that show paw and hoof prints. Each year the number of animals following such corridors has risen and now even grizzly bears, animals notoriously reluctant to go near well-traveled roads, use the Canadian passageways. Studies suggest, however, that some species that follow specific corridors, such as elk, will not go far beyond their established routes to use human-made crossings, which underscores the importance of studying wildlife migration thoroughly.

In the United States, where few such features have been built, an estimated 1.5 million elk and deer are killed by vehicles each year, with some two hundred human lives also lost in those collisions. Overall, it is estimated that one million vertebrate animals overall are killed *each day* by U.S. vehicles. But where wildlife-friendly overpasses or underpasses have been installed, collisions have dropped by as much as 83 percent.

My research also suggests that a problematic future faces the jaguar in the southeastern Mexico portions of La Selva Maya (the Maya jungle), a stronghold for the cat as recently as the 1970s. Here, as elsewhere, *Panthera onca* habitat is shrinking steadily. Population estimates varied widely in 2009, with knowledgeable scientists placing the total somewhere between 150 and 300 jaguars within the 1.8 million–acre Región de Calakmul, the Sian Ka'an biosphere, and three smaller adjacent reserves. These sanctuaries protect the largest remaining Mexican portion of La Selva Maya, which encompasses more than 5 million forested acres where Mexico, Guatemala, and Belize intersect. By comparison, research in the same time period held that no more than 200 jaguars roamed Belize's portion of La Selva Maya, with about the same number on the Guatemala side.

Mexico's approach to conservation in the region is distinctive. Rather than adopt a "hands-off" policy, the government's largest regional biospheres, Calakmul and Sian Ka'an, serve multiple purposes. The reserves are set up to shelter native flora and fauna while ensuring a relatively intact habitat to benefit researchers as well as places for local people to live and work.

The Mexican model makes innovative use of *ejidos,* a traditional form of multiple-purpose, cooperatively administered public land found throughout the nation. In this instance, authorities have set up doughnut-shaped districts around the boundaries of Calakmul and Sian Ka'an where settlers may grow crops, raise livestock, and build homes. In accordance with easement agreements, these residents may not hunt game or collect plants within the biospheres but are paid compensatory stipends and invited to participate in culture- and nature-oriented tourism. Locals find work as guides to wildlife and Maya ruins, for example, or selling handicrafts, lodging, and meals to visitors.

But population growth in the area verges on overwhelming. Since 1990 tens of thousands of new settlers—many of them Central Americans fleeing oppression and poverty—have relocated illegally to areas surrounding and inside the reserves. Most of these subsistence farmers follow a tradition of slash-and-burn agriculture that demands, as land fertility is exhausted, new patches of jungle every three to seven years. As is common throughout Latin America, some newcomers interpret a "biosphere reserve" as a de facto invitation to exploit—rather than to preserve—a natural resource. In addition, politically well-connected timber and cattle companies are eager to earn a profit from Mexico's last uncut jungles, and many do so with impunity.

In 1991, employing a strategy calculated to counter such pressures, an innovative program was launched by the Mexican nonprofit group Pro-Natura that has since been emulated elsewhere. Proponents reason that, given the inevitability that population growth will affect biospheres, the best strategy is to surround them with people educated to understand the value of environmental protection and given an economic incentive to defend and sustain it. The theory is that when villagers are convinced such a sanctuary is worth more to them intact than denuded, they may hold the line against further habitat destruction.

The approach, though not infallible, can boast of some success. Updated cooperative farming is weaning *campesinos* from slash-and-burn and shifting them toward more sustainable methods. Tillers of the land are instructed in organic techniques (as opposed to those that rely on chemical pesticides and fertilizers) and urged to plant climate-appropriate crops with high-nutrition yields. In designated areas, farmers are allowed to harvest certain game animals and wild foods in managed fashion. Specific medicinal plants can be taken, for instance, as well as honey produced by wild bees.

Before 2005 Mexico's southern nature reserves were also places where hunters were invited to shoot jaguars—not with bullets, but with tranquilizer darts and cameras. In 2001, for instance, the Pennsylvania big-game enthusiast Joe Bojalad paid upward of five thousand dollars for the chance to legally fire a rifle—three times—at a Calakmul jaguar. As described by Eric Niiler in *Scientific American* magazine, none of the ketamine-loaded darts hit the cat, treed by baying hounds. A local tracker eventually fired the shot that put the female into a short, tranquilizer-induced stupor. Over the next forty-five minutes a wildlife veterinarian and biology student examined, measured, and tagged the cat, then fit a radio collar to its neck, drew blood samples, and removed parasitic fly larvae that had burrowed into the animal's skin. (Parasites and other invertebrate pests not only irritate but can actually kill an otherwise healthy jaguar.) Their duties complete, the team crept away and watched the cat wake up and stagger into the forest on its own. It now wore a transmitting device that, if monitored properly, could help experts better understand Calakmul felids.

Capturing jaguars is notoriously difficult, yet more than twenty Mexican cats have been collared since the Calakmul effort began in 1995. Radio transmissions yield data that are combined with results from camera-trap photography, ground transects, hair snares, tracks, and scrapes. Such secondary evidence fills in key pieces missing from the jaguar puzzle.

"This is not just a scientific project for jaguars," said Carlos Manterola of the nonprofit Unidos para la Conservación, which cosponsored the effort with Mexico's national university. "We are using the jaguar as a flagship species for conservation of the Maya jungles."

Manterola's group brought ecotourists and writers, including the celebrated author Barbara Kingsolver, to Calakmul, along with recreational hunters and scientists. In part through fees paid by Joe Bojalad and other trigger-happy visitors, Unidos was able to pay locals cash compensation when jaguars killed their livestock, a strategy widely used in a bid to reduce poaching. The organization's educators also taught area residents how to forge a sustainable relationship with resident jaguars, while some funds collected from hunters and ecotourists went to finance scientific studies of the animals.

After 2005 the region's conservation management shifted largely to academic researchers and the auspices of Friends of Calakmul, a nonprofit group operating in both the United States and Mexico. Unidos para la Conservación expanded its activities to include northern Guatemala, where its darting program has continued on and off, with mixed results.

Overall, observed Roan McNab, director of the WCS program in Guatemala, "Mexican officials are doing a good job of managing their parks. They move homesteaders out, set up nearby *ejidos,* put military patrols on reserve perimeters, and get good research done. It's troubling, however, that Mexican settlers and ranchers are right up against the border of Calakmul." McNab's concern is that, human nature being what it is, people will inevitably sneak into a reserve and cause environmental harm or disturb corridors wildlife use to reach other habitat.

"The hope is that we can have a sustainable population of jaguars not only inside but outside the [Mexican] reserves," explained Gerardo Ceballos, an ecologist at Mexico City's Universidad Nacional Autónoma and codirector of the Calakmul project, in an e-mail interview. He conceded that some individuals in Mexico "make their living by illegally hunting jaguars. This represents a very serious risk for the conservation of the species."

Jaguar conservation in Mexico was elevated to the highest political level when then-president Vicente Fox declared 2005 to be "The Year of the Jaguar." During those twelve months, portions of the Sierra de Vallejo in the western state of Nayarit were decreed as "natural protected areas" specifically for jaguars. Similar designations were made in the mountainous states of Jalisco, Michoacán, and Oaxaca, where *Panthera onca* barely hangs on. Nationwide, thousands of acres became big-cat sanctuaries, at least according to official edict.

Yet in 2007 Ceballos described illegal hunting of jaguars as "rampant" and noted the breakup in 2002 of a single Mexican trafficking ring that yielded twenty-three jaguar pelts. In his country as a whole, the scientist pointed out, more than half of the original tropical forest cover is gone and an estimated one million additional acres are lost each year. "We may have won some environmental battles," Ceballos concluded, "but right now, we're losing the war."

秦

Some wildlife advocates are critical of the "hire a hunter" approach used in Mexico and Guatemala for jaguar darting. "I have a lot of problems with it," one longtime activist in the region, who asked not to be named, told me. "For one thing, a big cat that is bayed and then tranquilized can hurt itself if it falls sedated out of a tree. Also, I've heard that dead animals may be dragged around to help attract a jaguar toward a dart-hunter. That's simply unethical." (Elsewhere, large cats have been summoned—only to be shot with darts—through broadcasts over loudspeakers of recorded captive animals.)

Some scientists also worry about jaguars being tranquilized too often or getting dosed with too much or too little ketamine, a powerful drug whose dosage is largely determined in the field by guessing an animal's weight and health. A partially tranquilized cat may become disoriented and injure itself—or a bystander. Critics worry about the stress of the induced coma, which can lead to the death of a jaguar. Still others are morally opposed to having cats shot by overeager game hunters, who may not value conservation and have been known to shoot jaguars that are already collared.

Such collaring efforts are "being driven from a hunting perspective," chided Alan Rabinowitz, then of WCS, claiming that the Calakmul project's chief darter, Tony Rivera, talked openly about his desire to see limited jaguar hunting—with real bullets—return to Mexico. Rivera reportedly stated his belief that there are enough jaguars in parts of Mexico, including the Yucatán, to justify limited harvesting. "[That] is simply not true," said Rabinowitz, who has visited La Selva Maya. Gerardo Ceballos agreed, stating that while "there is a healthy, stable population of jaguars in Calakmul, we do not support jaguar hunting. . . . We do not have enough scientific information for that." He defended his employment of Rivera, however, despite the hunter's documented background as a poacher. "I am very happy," said Ceballos, "that he can make a living without hunting jaguars." The professor noted that Rivera and his associates lost their livelihoods when Mexico banned jaguar hunting in the 1980s and, thanks to conservation, have stable incomes from activities that are useful and legal.

Rivera in recent years recruited donor-darters through advertisements in *Safari News,* a publication catering to big-game hunters. Paying customers could spend a week stalking jaguars in Mesoamerican jungles. During 2006 the outfitter helped scientists find, dart, and collar jaguars in Guatemala, but these darting safaris were eliminated the following year, reportedly because the data collected were of minimal value relative to expense.

With Mexican and Central American strongholds losing ground, the region offering perhaps the best chance for long-term survival of jaguars may be the greater Amazon River basin of South America. The sheer size of its uncut rain forests and impenetrable wetlands attests to this. Uncounted hundreds—perhaps thousands—of big cats are believed to still roam this vast watershed, encompassed by parts of ten different countries. Brazil, the largest of these, has established wildlife protection areas as large as Belgium. Soon after taking office in 2003, President Inácio Lula da Silva ordered protection of new forest reserves covering an area bigger than Louisiana. Yet in Brazil, as in much of Latin America, the key to conservation success lies in enforcement rather than legislation and litigation. Critics insist that so-called protected areas are often merely undefended "paper" parks where loggers, ranchers, farmers, miners, and settlers operate with virtual impunity. Between these zones, natural corridors used by jaguars for millennia are disappearing.

According to statistics compiled by various entities, forest cover in Brazil has been vanishing at the rate of about ten thousand square miles a year since the early 1990s. Nurit Bensusan, coordinator of public policy for the World Wildlife Federation, told *E* magazine in September 2005 that soybean farmers and cattle ranchers are responsible for much of the Amazon rain forest's recent destruction. Soy and beef are key exports for Brazil. Bensusan said that comparatively little money has been used to enforce existing laws designed to curb illegal activities that destroy Brazil's natural resources.

"We're disturbed by the numbers," Ciro Gomes, minister of national integration, conceded when confronted with reports showing that Brazilian forest destruction reached its second-highest level ever between 2003 and 2004. "But the government," Gomes told a reporter, "is adopting a package of prevention measures that is changing this picture." Yet by 2008 little change could be seen. According to a January 27 report in the *New York Times,* Brazil's president that week announced emergency measures to halt the burning and cutting of the country's rain forests for agricultural use. In the previous five months alone, the Brazil government said, 1,250 square miles of rain forest were lost.

⇒⇐

From the jaguar's viewpoint, the closest thing to heaven might be the vast mosaic of swamps, savannas, and forests that sprawl across large portions of Brazil, Bolivia, and Paraguay. These areas continue to be rich in wildlife and low in human population. Jaguars here reach record-setting dimensions, weighing up to 350 pounds and measuring nine feet or more from nose to tail-tip. In the 74,000-square-mile Pantanal alone live as many as 15 percent of all remaining wild jaguars. But despite extensive research and great efforts to mitigate ongoing conflicts with ranchers, such jaguars are far from safe. A 2008 *New York Times* report stated that during a two-year period one Pantanal scientist, Fernando Azevedo, lost four of the fourteen jaguars he was studying to ranchers and cowboys. Among the casualties, according to Azevedo, were an adult female and her two cubs.

Reducing such killing is a top priority of conservationists as well as scientists, who fear that the cats cannot endure a high rate of poaching indefinitely. Biologists have presented research results suggesting that ranchers often exaggerate depredation by jaguars. Experts also acknowledge that cattle may actually enhance habitat for the cats' natural prey by opening up landscapes otherwise too dense with vegetation for the prey animals to flourish.

But too often, observers say, the response to a jaguar's mere presence by local residents in the Pantanal and Amazonia is simply to pick up a gun, set hunting hounds on a scent, and shoot to kill.

"We have to act," Sandra Cavalcanti, one of the foremost Brazilian biologists studying Pantanal jaguars, told the *Times*. Strategies advocated by Cavalcanti and some of her colleagues include instruction of residents in nonlethal forms of predator control and better livestock management, along with payments for jaguar-killed cattle and financial incentives for leaving the cats alone. Some ranches in Brazil have in fact cut depredation dramatically by stringing electric fences, deploying guard dogs, installing bright lights around pastures, instituting regular patrols, and even setting off fireworks at night. Concurrently, international groups such as the Panthera Foundation have bought large ranches and transformed them into jaguar sanctuaries, while some ranchers are protecting cats as part of sideline ecotourism businesses. All this may not be enough, however.

"We need a new generation to come along and change the old ways of thinking," Marcos Moraes, owner of the São Bento ranch, told the journalist J. Madeleine Nash, echoing sentiments expressed in other epicenters of human-jaguar conflict.

In neighboring Paraguay, jaguars have been harassed by Mennonite farmers, some of whom conservation-minded outsiders accuse of treating the animals as vermin. "Mennonites are very industrious people," said the biologist Anthony Novack, who has studied *Panthera onca* in that country as well as in Central America. "But they seem to be utterly unconcerned about the long-term environmental impacts that their agricultural activities have on the landscape."

Novack described in an interview a hundred thousand–acre hunting reserve and cattle ranch in Paraguay's Chaco district, near Filadelfia. "The

Mennonites gradually have cleared all of the land bordering that ranch, and although a few jaguars continued to hold on [until 2001], every one of them has now been killed. This is despite a healthy prey base inside the property."

Over a lunch of pizza in a California restaurant, Novack elaborated on the potential impacts of hunting on the foraging ecology of big cats, the subject of his master's thesis at the University of Florida. In a scat-based study in the Maya Biosphere of northeastern Guatemala, he found the jaguar diet dominated by armadillo, coati, deer, and peccary. "Law enforcement [concerning the hunting of jaguars] is generally non-existent or lax outside of the United States," Novack reported in his thesis, and "subsistence hunting [by humans] has depleted populations of major prey in areas lacking strict protection. Typically, as human populations increase, wildlife populations are depleted because hunter behavior is guided by efforts to maximize immediate harvest success and not long-term conservation." While Novack's research did not find any significant difference in the diets and prey selection patterns of jaguars in hunter-harvested versus nonhunted areas of forest, it raised other questions: "Reduced availability of large prey may result in reduced survival rates, larger home ranges, and, consequently, a lower carrying capacity" for jaguars in hunter-frequented areas. A separate study in Tikal National Park suggested that the relatively small number of jaguars found there—fewer than ten in 2006—could be a partial result of illegal hunting and a decline in prey animals.

Novack, who spent more than three years as a Peace Corps volunteer in Honduras, told me that his life experience as well as his research had led him to conclude that rural "communities must perceive some direct benefit from natural resource protection." If they don't, he said, long-term conservation in developing nations will continue to be difficult.

❧❧

One key to jaguars' long-term survival in Brazil, Paraguay, and elsewhere in Amazonia may be further refinement of research methods. In part because the cats are notoriously people-shy and occur in low densities, researchers have relied to a large extent on estimates extrapolated from camera-trap pictures. (Collaring is comparatively expensive, logistically challenging, and physically risky for both cats and researchers.) But some trackers following jaguars fitted with satellite-connected collars found that such cats, particularly males, roamed far more widely than previously thought. Unless this factor is accounted for, it may seem that there are more cats living in a "home range" than actually are, since some may be simply passing through. Thus some camera-trap research may lead to overestimates of population densities of jaguars by a factor of up to 100 percent. In the Pantanal, for example, some extrapolations were slashed. One researcher cut his figures from 11.7 to 6.7 jaguars per thirty-nine square miles (one hundred square kilometers). Meanwhile, in parallel research that may be relevant to jaguars, at least one study involving the camera trapping of tigers in India suggested that some cats may become "trap shy" and avoid such installations, and that inadequate spacing of the units may also skew results negatively.

Enlightened and creative sanctuary management could become another key source of hope for the jaguar's future. Increasingly, indigenous people whose ancestors have a long relationship with these cats are being called upon to help save them. In southeastern Bolivia, representatives of Fundación Ivi-Iyambae, Captinía de Alto, y Bajo Isoso help manage one the world's great jaguar strongholds, Kaa-Iya Gran Chaco National Park. When established in 1995, it was the only park of its kind in the Americas established and administered by indigenous people. The dry tropical forest, at 8.4 million acres encompassing an area larger than Massachusetts and Connecticut combined, was estimated in 2008 to hold as many as a thousand jaguars. It also is home to giant armadillos, Chacoan peccaries, and Chacoan guanacos, of which only an estimated 140 remain in Bolivia. Using monitoring techniques developed originally to determine tiger populations in South Asia, researchers estimated the Bolivian park—the second-largest park in all of South America—had about one jaguar per thirteen square miles. They warned, however, that poaching, illegal settlement, and deforestation were taking a significant toll on these felids.

CHAPTER TWENTY

"The Mother Liquor from Which We Have Come"

SATISFIED THAT NO HUMANS ARE NEAR, the jaguar sheds the forest's gloomy cloak and pads silently to the water's edge. The damp night is moonless, with a blur of gray-blue haze suspended like smoke above the tree canopy, drifting on a barely perceptible breeze. The musty smell of organic decay pervades the air, mixing with the sweet perfume of orchid and jasmine. Drops of moisture drip from dew-laden leaves. The cat sniffs for a moment, pauses, and pricks up his black-lined ivory ears. They pivot back, forward, and sideways. He is evaluating carefully the nocturnal orchestra of insects and amphibians, a cacophony of hums, clicks, shrieks, pings, and buzzes. A jungle partridge sounds its plaintive call, seemingly answered by a tree frog. Wide, unblinking eyes scan the water, sweeping the hazy horizon from left to right. Someone watching might assume that the motionless animal is staring blankly, lapsed in a daze. On the contrary, this felid brain is intensely busy, evaluating every sensory input. Time ticks by. The jaguar concludes that there is nothing out of the ordinary. For now, his route is clear.

It will take nearly half an hour for the young male to swim to the opposite bank. There are no significant obstructions or currents, but the channel is deep and busy with traffic around the clock. Breaks like this are fleeting. A fast, steel-hulled ship could appear at any moment. A quick transit is important. The jaguar is confident of making it to the other side. A strong, accomplished swimmer, he is not worried by physical demands. This cat knows that he is vulnerable in the water, from which any escape could be fatally slow. He does not like being in the open, where risk of discovery is high.

But the imperative pushing him to find and occupy new territory will not be denied. If he does not attempt this crossing, the feline knows that his current circumstances are precarious, and could lead ultimately to his death. The south bank is crowded with older males, each willing to fight to defend a chosen homeland and potential mate. As a barely adult three-year-old that has never fought another cat—discounting playful tiffs with his mother and siblings—the departing jaguar must find an unchallenged piece of jungle that can support him with plenty of wild game and ambush cover. He glances left and right one last time, plunges into the lukewarm water, and begins paddling with his oversize feet toward the opposite bank.

I consider myself a fair swimmer, yet the prospect of a freestyle traverse of the Panama Canal is daunting. In my youth I may have had the gumption and ability to make the crossing, but negotiating with an unfriendly ocean-going vessel is not my idea of a good time.

Fantasy scenarios starring man and beast unspool in my mind as I stand beside the Miraflores Locks on a sultry, thunder-rocked afternoon. The annual wet season has surely begun—with a vengeance. I behold a spectacle that, like the Grand Canyon and northern lights, is so visually arresting that the tools of descriptive language are abysmally inadequate. A hundred yards in front of me, drenched by sheets of rain, bobs a full-size oil tanker, the largest type of nonleisure vessel plying the high seas. Tons of encapsulated mideastern crude descend slowly as water pours in from the Pacific side of the lock. Only a few inches of clearance—barely enough room to stuff a big-city telephone directory—lie between the ship's hull and the steep-sided compartment. Canal workers busy themselves uncoupling cables and barking commands. An electric train engine called a mule eases the tanker along its journey from Saudi Arabian dunes to California refineries.

A few yards beyond the tanker's rudder, in the adjacent lock, floats a second behemoth: a Chinese merchant vessel, possibly heading to Shanghai from Manhattan. Nothing has prepared me for the sheer enormity of this freighter, longer and taller than an NFL football stadium. The titan's open

Figure 20.1. Jaguars, which often inhabit wetlands and rain forests, are excellent swimmers and fond of water, as this cat demonstrates. The animals will swim across rivers, canals, and lakes in search of food and new territory. Photo by Julie Larsen Maher, © Wildlife Conservation Society

deck is stacked 150 feet high with the sealed containers mariners call TEUs (technically, twenty-foot equivalent units, but sometimes applied to the containers regardless of size). Each metal box measures forty feet long by eight feet wide and nearly nine feet high. This load is nineteen containers deep and could hold as much as $300 million worth of cargo. I wave at the pilot, as insignificant as a flea, peeking from his glass-walled helm atop a twenty-story navigation tower. The throbbing shudder of the ship's idled engine, entrained with that of the tanker, reverberates along the canal like a ragged, bass-note reggae pulse.

Above the bristling antennae and spinning radar dishes, a steamy panorama of glistening emerald looms incongruously beyond the canal. The rain forest looks like nothing so much as a painted backdrop in a Hollywood movie. *Jurassic Park* comes to mind. The thunderstorm still rages overhead, like the tympani section of an orchestra. I compare the lightning bolts to crashing brass cymbals, the heaven-sent rhythms to pounding copper kettledrums.

I have arrived at a place where the unquenchable thirst of consumerism meets the timeless life cycle of the fecund tropics. At this nexus the denizens of a nearly pristine Central American jungle can peer into the mechanized world of *Homo sapiens*. I wonder how felid minds—which must have witnessed some version of this scene a thousand times—regard the Miraflores Locks. How do jaguars react, if at all, when they stumble upon a ship loaded with Arabian petroleum or Chinese TV sets?

"I haven't a clue," a retired Canal Zone security guard tells me over an evening meal, when I pose the question a week later. "I would come across jaguars on my patrols, but it was really very seldom. Pumas, yes; *tigres,* no. The jaguars really don't *want* to be found. But, *por su puesto,* they are definitely out there. Yes, even around the locks where you were standing."

I ponder Roberto's provocative reply. In the background, my friend Daniel is telling our other companions a ribald joke. From a building next to the restaurant come the raucous sounds of a crowd gathering for a Saturday night cockfight. We sit in candlelight, accepting the inevitable Latin American power outage. My new acquaintance pauses and smiles, apparently recalling with fondness his late-night encounters with big cats, some of which almost certainly paddled across the man-made channel.

"Incidentally," Roberto says at last, "did you know a guy once swam the entire length of the Canal, including all of the locks?"

I nod. "Yeah, I read Richard Halliburton's story in the museum. It's astonishing."

Roberto nods, lighting a cigarette. "He paid a thirty-six-cent toll; the lowest ever. But it was a difficult swim, and it took Halliburton eight days."

People rarely see a jaguar paddle across the Panama Canal, but ship captains, maintenance workers, and passengers swear that it happens. This is reassuring to wildlife biologists.

"No one knows how many jaguars travel between Central and South America," Costa Rica's Eduardo Carrillo told an interviewer, "but it is vital to the health of this species for a habitat corridor to remain open across the narrow waist of Panama. That's why we study *Panthera onca* on both sides of the isthmus."

Fortunately, a north-south passage for jaguars continues to exist through Panama, although it is increasingly fragmented by agriculture, hunting, and urbanization. Thousands of cars and trucks shuttle daily, for example, between the bustling Atlantic cargo port of Colón and the commercial center of Panama City on the Pacific. Vehicles ply a modern highway that runs roughly parallel to the Panama Canal and its companion railroad. These are busy thoroughfares capable of intimidating any animal that attempts to cross them.

The juxtaposition of natural and unnatural yields some profound ironies. At the narrowest pinch of the isthmus, jaguar habitat is actually *shielded* from development by the canal. Unmolested forest protects a vast watershed that stores and releases the millions of gallons of water needed each day to float ships and operate locks. Streams and springs trickling from this thick jungle also supply nearby cities with plentiful fresh water. Thanks to this sanctuary, the rugged spine of Panama looks much as it did when French crews, followed by those of the Americans, excavated the big trench during the late nineteenth and early twentieth centuries. Taken as a whole, the flora and fauna found here remain more diverse than almost anywhere on our planet.

This abundance prompted the naturalist David Fairchild, writing in a 1922 issue of *National Geographic*, to laud Panama as a place that may still look like it did before humans existed. In this jungle Fairchild found an environment "where life teems and new forms develop, in the midst of that living stuff up out of which man came ages ago." Such tropic ecosystems, he contended, "are the mother liquor, so to speak, from which the plants [and animals] of the world have come."

Generations later, field biologists tell me that we are in a race against time, not only in studying and saving the jaguar but also in understanding its rain forest home. They note that scientists did not begin to evaluate the tropics in a serious way until around 1800. A century later, specimens of even large mammals were being collected for the first time. Elusive animals took even longer. Jaguars received scant attention from biologists before 1970. Compared with research in Earth's temperate zones, our knowledge base for equatorial environments is in its infancy. Yet the small amount we know already has life-altering significance. The warm, wet ecosystems of the world are an incomparable source of food, fiber, energy, genetic information, clean air, pure water, life-form diversity, waste decomposition, flood abatement, drought mitigation, erosion control, and crop pollination. They also constitute a priceless resource for healing the human body, inasmuch as an estimated 40 percent of our medicines are believed to devolve from compounds obtained originally from tropical plants.

Panamanians assured me that jaguars were still being seen in their nation, albeit less frequently than ever before. My original intent was to travel to the thick jungle of the Darién Gap, along the border with Colombia, but I gave up the idea when several informants—including a Christian missionary and an ecotourism guide, both of whom had been in the Darién many times—convinced me that it was too dangerous for a casual, nearly penniless visitor. Armed escorts were needed, the Panamanian guide told me, owing to ongoing violent attacks in the region related to long-running civil unrest. Men carrying automatic rifles guarded the few biologists who dared to work in the area. Alan Rabinowitz and Eduardo Carrillo had trekked into the Darién a few years earlier, with hired guns at the ready. While this dodgy situation might help protect jaguars to some degree from rampant poaching and forest destruction, habitat loss in "safer" areas of the Darién and across the border in Colombia suggested that the cats were being squeezed into ever-smaller territories. Scientists were trying to determine how many jaguars—if any—still traveled through the gap between Central and South America.

When my quest brought me from New Mexico to the slender arm linking North and South America, I was still determined to see a wild jaguar. Yet during my ten days in Panama something shifted. It happened a few days after my visit to Miraflores, as I made one last daylong hike into the rain forest in the hope of glimpsing *Panthera onca*. Because time and transportation were limited, I chose a swath of jungle immediately north of Panama City. This protected area is said to be the only rain forest in the world contained within the municipal boundaries of a major city. Civilization loomed so close to the park that the towers of international banks and luxury condominiums were visible from the crests of the park's hills. Yet despite its proximity to a human landscape, advanced biological research was under way here by the Smithsonian Institution and various universities. Prey animals were abundant, and I had reliable evidence that jaguars still inhabited the sanctuary. On a weekday morning I took a bus to Panama City's main terminal and hopped a taxi to the park's visitor center. A friendly warden handed me a map and pointed me in the right direction.

About an hour into my hike, I paused beneath a tall ficus tree. Looking skyward into its torso-thick limbs, I noticed a troupe of spider monkeys feasting on ripening fruit. Their high-pitched chatter and rambunctious antics disturbed an algae-covered sloth, which crawled with arthritic deliberation across liana vines to a neighboring custard apple tree. I sat on a decaying stump, sipped from a water bottle, and watched the show, musing idly about my odyssey.

I had covered thousands of miles and spent thousands of dollars during the twenty-seven months of my off-and-on adventure, yet the ever elusive, supremely adaptable jaguar had not revealed itself. The closest I had come was seeing various tracks, scrapes, and other physical signs; hearing a full-throated roar in the small hours of a full-moon night; and communing with a couple of edgy "problem" jaguars recently removed from the wild. I had studied scores of camera-trap photos and seen a dozen generally lethargic jaguars displayed in zoos. That was it.

The curious thing was, rather than being disappointed, I now felt sanguine and content—almost blasé. This sense of equanimity surprised me. It was so unexpected as to seem almost funny. Instead of laughing, however, I felt a wave of enormous relief sweep through my body like the warm belt from a shot of whiskey. Instead of being frustrated by the cat's evasiveness, I now regarded *Panthera onca* with heightened respect and admiration.

A couple of lines floated back from Peter Matthiessen's book about his quixotic search, undertaken with his fellow naturalist George Schaller, for the Himalayas' rare and secretive snow leopard. "I am disappointed," wrote Matthiessen, about not seeing the cat, "and also I am not disappointed. That the snow leopard is, that it is here, that its frosty eyes watch us from the mountain—that is enough."

I was resigned to what my lack of "success" said about jaguars—and me. In my hubris, I had assumed I could outwit a highly evolved alpha predator that had been evading all conceivable enemies for hundreds of thousands of years. The jaguar was damned good at the game, because it had to be. I was not, because my survival did not depend on it. My rudimentary outdoor skills, which had guided me through hundreds of miles of backcountry and once helped me survive ninety-seven days alone in a frigid alpine winter, were no match for this sly, smart cat. You cannot plan to see a jaguar, I now realized, any more than you can plan to have a religious experience or meet the love of your life. Conditions can be created to heighten probability, but chances of realization remain slim.

"We've seen so much," Schaller tells Matthiessen at the end of *Search for the Snow Leopard*. "Maybe it's better if there are some things we *don't* see."

I felt no regret about my own choices vis-à-vis the jaguar. I had made my way—largely alone—through forests in seven countries without benefit of tracking hounds, hunting guides, infrared goggles, GPS units, cell phones, satellite photography, deceptive scents, live bait, or imitative jaguar "callers." Was it unexpected that one of the world's most aloof felids had not stepped forward for an introduction? Why should it?

This chain of reverie underscored another ironic twist in my shaggy-dog story. The Panamanian jungle where I sat was teeming with life. Everything in sight seemed viscerally alive with cell division and fertility, growth and decomposition. Whatever I saw around me, all that I could touch, was a product of nature. Thanks to the jaguar, I'd borne thrilling witness to this constantly unfolding marvel for several years. Yet the inspiration to write the book you are reading had been sparked in the *most* unnatural place I could imagine: New York City's Times Square. On June 17, 2003, on an infrequent visit to Manhattan, I was standing in a long, slow-moving line at a discount ticket outlet, waiting to secure two seats at one of that evening's Broadway shows. I bought a newspaper to pass the time. Browsing through that morning's edition of the *New York Times,* I read an article about jaguars that quoted Costa Rica's foremost jaguar expert, Eduardo Carrillo. This, as much as the 1996 *New Mexican* story about Warner Glenn, had fueled the fire.

Each time he saw a jaguar, Carrillo told reporter Natalie Angier, it felt "like a miracle or a dream, the most exciting thing you can imagine." The wildlife biologist told Angier that while he had seen jaguars in the wild on at least two dozen occasions, such cats had probably seen *him* scores, perhaps even hundreds, of other times. It suddenly made sense that jaguars had seen me, too, without my knowledge. Bravo for them, I thought. And lucky for me.

❧

The playful monkeys migrated to another tree, their progress marked by the shrill shrieks of parrots and parakeets. Primates invariably upset the avian community, as do other meat-eating mammals. Gregarious birds set up a ruckus when a jaguar walks below them, a partial explanation for the felid's nighttime stalking. *Panthera onca* was not putting in an appearance this Thursday afternoon in Panama—though who knows what might be peering through the underbrush.

My mind stayed adrift, lapsing into currents of memory and eddies of emotion. I reflected on desert rats Warner Glenn and Jack Childs, scrabbling through the desiccated mountains of New Mexico and Arizona. I pictured them stumbling unexpectedly on jaguars in unlikely places, looking for one animal and finding another. In doing so, these men—and their companions—had joined the exclusive club that comprises those who have found creatures either previously unknown or believed extinct or extirpated. The big-cat biologist Alan Rabinowitz qualified for membership in 1997 when he found Burmese villagers hunting a new-to-science leaf deer, so named because it was tiny enough to be wrapped in a single leaf from a local plant. Other lucky explorers had chanced upon the Congo's okapi in 1901, Japan's Iriomote cat in 1967, Paraguay's Chacoan peccary in 1972, the Vietnamese antelope in 1993, and, in 2005, an entirely new species of primate in Tanzania called a kipunji. A similar mix of amazement and glee must have invigorated Arkansas researchers who believed in 2004 that they had found the ivory-billed woodpecker, more than sixty years after its presumed disappearance from swampy forests of the American South.

The jungle's rowdy monkeys and squawking parrots gradually disappeared, and the bush regained its groggy midday composure. The air was thickening now, saturated with moisture and dulled by the numbing drone of cicadas. Rain would come within the hour. I watched the lethargic sloth inch back to its favored perch and settle down for a postlunch siesta. The forest grew eerily silent. The only creatures now stirring were leaf cutter ants, marching purposefully along circuitous pathways, carrying bits of green tissue to their underground cities for mastication and storage.

My mind wandered again. I was in a reflective, bittersweet mood. My nine-week journey through Central America was nearly over. Within days, I would be back home in the redwood forests of northern California. I sat without stirring in a welcoming pool of silence and solitude. After several minutes, I removed from my pocket and polished a carved amulet I had bought from an artisan in Maya Center. The black slate, about the diameter of a dollar coin and thickness of a child's finger, revealed the stylized face of a jaguar. Its lines were carefully etched: a flat triangular nose, oval eyes, scalloped ears, long drooping whiskers, and irregular rosettes. I had carried the slate carving everywhere for a month in the hope that it would help lead me to a flesh-and-blood jaguar.

I tucked this talisman away for the last time. My search was over.

What had it meant? I began to understand that what enchanted me about jaguars was not what they were, exactly, but what I believed they represented. For me, *Panthera onca* embodied the transcendent quality of Earth's deepest mysteries, of all in the natural world that is still pure, wild, and unmolested. Yet I knew that even this was illusory, since human beings have managed to affect so much of our planet. No place on Earth seemed untouched by the upright, bipedal toolmaker with the big brain and opposable thumb. Was there any purity left in Nature?

"In the end," Grahame Webb, a pessimistic (yet perhaps brutally realistic) Australian crocodile expert, told the nature writer David Quammen in *Monster of God,* "people are not gonna conserve something that has no use or value—that, in fact, has a negative use or value. It's not gonna happen. It never has, in the history of the world."

By now I was convinced that no creature in the Western Hemisphere has meant so much to so many for so long as the jaguar. I dared to speculate that, because of this enduring relationship, my fellow humans might be less likely to give up on—or spurn—this vanishing spotted cat. I knew I could be wrong. In a highly mediated, consumption-oriented society, perhaps the flesh-and-blood jaguar had lost the special values attributed to it for millennia. The chain of pre-Columbian cultural heritage that linked Venezuelan cowboys, Maya rulers, Amazon shamans, Hopi hunters, South American fur traders, and Aztec warriors already could be broken. Maybe the luxury cars, the football teams, and other iconic "borrowings" were now more important than preservation of the real thing. It was a sad and disturbing thought, but it rang of truth.

I came to a conclusion that should have been obvious to me from the beginning. My task was a bigger challenge than I was willing and able to accept. I had been reluctant from the outset to plunge deeply, without hesitation, into the most physically challenging of habitats, where the wild jaguar is most at home. Such locations are often uncomfortable and dangerous. Was it any surprise that Alan Rabinowitz, even in his fifties, was a musclebound weightlifter and martial arts enthusiast? (Even after several years of debilitating cancer treatments, the zoologist had not let disease slow him down.) I now also accepted that jaguar strongholds can be expensive to reach and often demand the services of a knowledgeable escort. I had been unwilling—and unable—to make such investments. Instead, I had skimmed the surface, compromising my ease only for a few days at a time. I had eschewed tracking dogs, knowing all along that even if I had used them my odds of spotting *Panthera onca* were abysmally low. Wendy Glenn, Warner's wife, reminded me that her husband "only saw two jaguars in seventy years of near-daily looking, much of that time with very well-trained hounds."

Wendy was referring to an unparalleled stroke of luck. Warner and his hounds had treed a second jaguar almost exactly ten years after finding the first. The occasion was a sanctioned "predator control" hunt for a mountain lion on a private ranch in New Mexico's Animas Mountains, about twenty miles east of the Peloncillos. On February 20, 2006, Warner and his daughter, Kelly, were following a scent when one of their hounds went missing. But Powder soon reappeared, nursing a fresh injury in his neck and shoulder.

"Something had whipped the dickens out of him," the then-seventy-year-old guide recalled. Members of his mule-mounted party were concerned the "something" might be a big javelina, which can mortally wound a dog. Following their frantically barking hounds, the pursuit team brought to bay a middle-aged, two hundred–pound male jaguar in a juniper tree. "My gosh," shouted Warner, who proceeded to snap pictures of the snarling cat, which he decided to call Border King. He was the fourth member of his species confirmed in the United States within a decade. This jaguar was not reported seen again, though Warner believes he and others may very well be out there on the northern side of the boundary line.

Besides enlisting the help of an expert if I wished to encounter a jaguar, I now believed I would need to explore what the novelist Joseph Conrad famously referred to as "the heart of darkness." This is the unfamiliar psychological territory where all bets are off, laws are arbitrarily enforced, and the usual assumptions do not apply. Anything short of a plunge into this shadowy unknown—the domain of black-market wildlife poachers,

Figure 20.2. Warner Glenn nicknamed this adult male jaguar Border King after encountering him in the Animas Mountains of New Mexico on February 20, 2006. Photo © Warner Glenn

looters of archaeological treasures, and big-game hunters aching for illicit trophies—would bring me only the slimmest opportunity of meeting the Maya god of night and guardian of the underworld. Playing it safe made it harder to find a creature that did not want to be found. For the first time, it struck me that maybe I had not really *wanted* to succeed. Perhaps I had been rooting all along for the jaguar to win this sport of hide-and-seek.

I pondered a final unsettling reality: that deep down I preferred to leave the world's last wild jaguars alone rather than disturb them. While I felt deeply that humans ought to cultivate conditions to promote the cats' survival and to study them, it was probably best for both species if they remained essentially apart. In order for people and jaguars to survive, their lives should move as parallel lines: going forward, but not touching. If allowed to exist in their respective realms, *Homo sapiens* and *Panthera onca* had less chance of disturbing one another.

Lest this sound like an unconvincing rationalization, I also knew that if I somehow came upon a wild jaguar I would feel ecstatic. My heart would swell with the elation of a dream fulfilled. What had changed was my belief that a continuation of my search made sense. Knowing what I knew now, staying the course seemed beyond imprudent and bordering on foolishness.

CHAPTER TWENTY-ONE

"To See One at All Is a Lifetime Experience"

MY EMOTIONS AS MY PLANE LIFTED off from Panama were unsettled. I felt as troubled by the prospect of a world without wild jaguars as I might be if the human race were to abandon art or extinguish mystery—or the pursuit of religious ritual, or the conundrum of paradox, or the splendid thrill of astonishment. Throughout the New World, jaguars had played a starring role in the staging of such compelling experiences. They were an important part of what made us human. For a shrinking number of indigenous tribal people, *Panthera onca* still represented not merely supremacy, prowess, and freedom, but the distillation of life itself. The exotic cat continued to serve as their companion, totem, or deity. If nothing else, disappearance of this felid would end one of the greatest ongoing dramas the Americas have ever known.

But in the modern world of cities, television, air travel, and the Internet, the jaguar is reborn as a new sort of symbol. Bordering on caricature, it has been chosen to represent the forests and swamps we left behind—and preservation of what remains of the Western Hemisphere's contested wilderness. Protecting jaguars has become an easier "sell," to borrow marketing parlance, in countries that the cats still call home. Thankfully, many who hear these earnest pleas do contribute to efforts on the animal's behalf. In the back of this book I list contact information for a few of the many worthy groups that champion the cause of jaguar preservation.

By saving top-of-the-food-chain animals such as big cats, so the conservation message goes, we save what sustains them. Remove a keystone species from an ecosystem, we are told, and the natural foundation of life starts to crumble. It seems that the flora and fauna of the Americas upon which jaguars depend are interwoven in a web of codependency. In the long-term view of many scientists, *Panthera onca*'s slow fade does not bode well for other terrestrial species—including our own.

Throughout Latin America, the tug-of-war between preserving the old and building the new seems never-ending. Sometimes it is hard to tell which side is winning. In Panama, for example, I was surprised to find that a tradition of preservation persists despite many stresses facing wildlife. Such hopeful signs, however, inevitably seem balanced by darker trends.

In her 2001 book *The Tapir's Morning Bath,* the science writer Elizabeth Royte noted that after more than sixty years of vigorous protection, a research station overseen by the Smithsonian Institution was still under constant siege. (This living laboratory, once a forested hilltop, was now an island located in a lake created by an impoundment for the Panama Canal.) Royte reported that during her months-long stay on Barro Colorado, armed guards still caught as many as three poachers each night. Sadly, trespassers had eliminated all resident jaguars from the island many years earlier.

In the absence of large predators, the collared peccary and coati greatly increased their population and rose to the top of the island's food chain. Seed-eating mammals, notably the paca and agouti, also became much more abundant. This, in turn, negatively affected the reproduction of certain plants and the survival rate of ground-dwelling birds. The absence of jaguars and mountain lions thus had cascading effects on Barro Colorado's ecology, leading to dramatic changes in fauna as small as insects and flora as large as high-canopy fruit trees. Similar changes have been noted in other places where large carnivores were eliminated, including many in the United States.

"People here are fooling themselves if they think their work is related to conservation," one pragmatic scientist working on Barro Colorado told Royte. "Most of it isn't. If they really wanted to conserve habitat, they'd work to redistribute the wealth and get slash-and-burn farmers, ranchers, and loggers to stay out of the forest. Conservation is a political problem."

In Latin America, as elsewhere, protection guarantees tend to evaporate when pitted against projects deemed to have greater, more immediate value than research, biodiversity, or beauty. Advocates of environmental protection compete against the human desire for jobs, homes, water, food, land, money, and other resources. Global warming and economic instability have complicated the situation even more. In Costa Rica the rainy season came a couple of weeks late in 2007 and reservoirs dried up. In a country where more than 80 percent of electricity is generated by hydroelectric dams, this meant blackouts and shutdowns. Because its climate seems to be getting hotter and drier, Costa Rica is now considering construction of new dams in national parks, which have been off-limits so far.

Still, miracles of nature continue to happen, even in places like fast-developing Panama. In 1992 a Smithsonian bee researcher reported seeing a jaguar cub playing along Pipeline Road in the Canal Zone town of Gamboa. "Jaguars are one of the most endangered animals in the world [and] to see one at all is a lifetime experience," wrote Nicola Smythe in a scientists' newsletter, as reported by Royte. "Yet here we have good evidence that jaguars are breeding within an hour's drive of Panama City."

Some of those who are inspired by this recurring ecological linkage pursue difficult, frustrating, and unsung work on behalf of the elusive jaguar. Why? I don't believe it is simply because they recognize the cat's important ecological role as an alpha species. After meeting many of these dedicated individuals, I have come to consider additional possibilities. I believe that some do their work simply because such an animal has a right to breathe the same air we do; because it is wondrous and beautiful; because it is highly intelligent, fearsome, and mysterious; and because a planet *with* jaguars is infinitely richer than one without them.

On the crowded Pacific island republic of Taiwan, the Formosan clouded leopard is regarded as a stunningly lovely and fabled creature. This handsome cat, a relative of other clouded leopards that are fast disappearing across mainland Southeast Asia, derives its name from delicate, irregular markings on its flanks said to resemble gathering storm clouds. There have been no confirmed field sightings of the reclusive subspecies for

Figure 21.1. The biologist and Belize Zoo founding director Sharon Matola feeds Junior, a cub born shortly after the capture of its mother, designated a "problem jaguar" by authorities. The mother rejected Junior, a cat that will of necessity spend its life in captivity. Photo courtesy Belize Zoo

decades, leading some experts to consider it extinct in the wild. But this has not extinguished flames of hope that burn in the hearts of Taiwanese people.

At the behest of the island's government, Alan Rabinowitz searched in vain for the Formosan clouded leopard but concluded that it may still exist in the wild. Chang Wan-fun, a researcher at Tunghai University's Department of Environmental Science, found evidence to support this notion in 1990, when he discovered what appeared to be a dead cub in a hunter's trap.

The Formosan clouded leopard's existence—and survival—is said to depend on whether sufficient prey animals occur in the mountainous Shei-pa National Park, established in part at the behest of Rabinowitz in one of Taiwan's last intact forests. This kind of cooperative effort underscores the importance of trying to save an apex species by protecting ecosystems in which it thrives.

❦

Just as the people of Taiwan hope their beloved leopard will reappear, ordinary citizens in highly developed, nontropical places are determined to document the presence of charismatic cats on their own homogenized (and presumably tamed) landscapes. The jaguar, it seems, is one among several large felid species people claim to have come across in the most unlikely of places.

In the British Isles, as improbable as it sounds, rural residents claim to have seen scores of exotic cats, including jaguars (often called panthers or jungle cats in the United Kingdom). As reported by Ron Toft in London's *Financial Times,* the Dartmoor-based British Big Cat Society counted 2,123 sightings of large felids in the United Kingdom from April 2004 through July 2005. This roster included mountain lions, tigers, leopards, and lynxes as well as jaguars. At least 17 reports included cats with cubs.

"There are photos and video footage and physical evidence in the form of animal kills and ground and tree marks," according to Toft. "A few big cats have been caught and some others have been shot or found dead." Some exotics—including a jaguar in Shropshire—were reported hit by cars and trucks. Others are said to have attacked horses, cows, sheep, and family pets. The society believes that these large carnivores probably were released intentionally, escaped from captivity, or are the offspring of such animals. Most have been sighted in the southwest districts of Dorset, Cornwall, and Somerset. (It should be noted that physical and photographic evidence cited to back up these claims has not held up well under the bright light of scientific expertise.)

Halfway around the world, in July 2006, the Australian Broadcasting Corporation reported some "220 sightings of a big cat in the western Sydney suburbs around Hawkesbury." The network said "a large number of domestic animals, including goats and dogs, have been savaged in the area." The newscast quoted the University of West Sydney scientist Rob Close as saying that the size of pugmarks gave strong evidence that the Australian marauder was truly a large (and hungry) exotic felid. Added the local resident Michael Williams: "We're dealing with an animal that's eating out 20 to 40 kilograms in a sitting."

In December 2007 authorities in Barfield, Tennessee, received a call from an anxious resident who was convinced that a "black panther" was prowling near Barfield School. Police investigated and found no such cat, but the report prompted two other Tennessee observers to volunteer that they, too, had seen pantherish creatures near their homes in previous years. "I was positive of what I saw," insisted a Rockvale resident, as reported by WGNS Radio. "[It was] an extremely fast, black, cat-like animal with the longest black tail you could imagine."

Many similar reports have come my way from around the United States since I began my jaguar research. If one expands the category to include mountain lions, there are literally hundreds of Americans who believe that they have seen big cats where there are supposed to be none.

What are we to make of these seemingly outlandish reports—and others like them turned in by people all over the civilized world? Perhaps they represent no more than a bored public's wishful thinking or a poor understanding of science. Could those making these sightings be mentally unstable, be addled by drugs, or have too much time on their hands? Maybe the cats really are escapees on the lam from zoos, wildlife parks, and private menageries. After all, thousands of large exotic felids are owned by ordinary citizens and often kept under minimal security in their backyards.

I believe it also is possible that people claim to see jaguars (and their felid cousins) because they do not want to imagine our planet without them. Maybe we simply are not ready to give up this lovely and evocative vestige of wild-ness and wilderness. But one thing is clear: conflicts between people and jaguars, as well as other large cats, must be resolved satisfactorily if such felids are to continue roaming the Earth freely. Only humans know this. The animals themselves have no clue how dangerous we are. Though many jaguars, like other felines, are curious, they did not evolve to care much about *Homo sapiens*.

"In an evolutionarily abrupt turning of the tables," noted a panel of biologists in *Large Carnivores and the Conservation of Biodiversity,* "humans are now responsible for the survival of large carnivorous animals." This empirical conclusion underscores a mythic belief among an indigenous tribe in Colombia that "the jaguar was sent to the world as a test of the will and integrity of the first humans."

Given current trends, jaguars may be doomed to eventual extinction outside captivity. At the end of 2007, according to the World Wildlife Federation, the jaguar population in Argentina, Paraguay, and Brazil was one-sixth to one-seventh what it was in 1990. Only the heart of the Amazon region provided a safe haven. Sadly, total eradication of *Panthera onca* before the twenty-second century is a real possibility. The science writer David Quammen suggested in 2003 that 2150 would probably be the endpoint for the special relationship between humans and wild predators. Such a disappearance would further isolate us from a natural world that sustains us, despite our attempts to manipulate, dominate, and destroy it. If nothing else, the jaguar's demise will deny our descendants the marvel of beholding one of Earth's most superbly designed creatures. It is a highly evolved animal that requires freedom in order to exist. In order to survive, it seems, the jaguar's independence and autonomy now must be defended with passion and aggression against the many powerful forces marshaled against it.

1. "GOD ALMIGHTY, THAT'S A JAGUAR!"

Alan Rabinowitz has been an authority on jaguar conservation since the mid-1980s, and I quote him frequently. Sources include e-mail correspondence as well as his many books, magazine articles, media interviews, lectures, and writings in scientific journals. The first quotation in this chapter is from *People and Wildlife: Conflict or Coexistence?* ed. Rosie Woodroffe, Simon Thirgood, and Alan Rabinowitz (New York: Cambridge University Press, 2005), 278–285. I had e-mail correspondence with Eduardo Carrillo in March 2006 and January 2007. The quotation attributed to him here is from his "Tracking the Elusive Jaguar," *Natural History,* May 2007, 30–34.

I interviewed Warner Glenn on several occasions between 2004 and 2007, via letter, e-mail, and in person. I use quotations and information from Glenn's *Eyes of Fire: Encounter with a Borderlands Jaguar* (El Paso, TX: Printing Corner Press, 1996) and from articles by Jeremy Kahn ("On the Prowl," *Smithsonian,* November 2007, 85–92); William Stolzenburg ("Species and their Saviors," *Nature Conservancy,* May 2001, 40–47); Peter Friederici ("Return of the Jaguar," *National Wildlife,* June–July 1998, 48–51); and Susan McGrath ("Top Cat," *Audubon,* August 2004, 48–55). I also drew information from Barbara Ferry's April 2, 2004, "Borderlands Jaguar" segment of *Living on Earth,* a public radio series (www.loe.org/shows/shows.htm?programID=04-P13 -00014#feature4). The Ambrose Bierce quotation is from *The Devil's Dictionary* (New York: Oxford University Press, 1999).

For the basic natural history of jaguars, I relied on many sources. Primary were Douglas H. Chadwick, "Jaguars: Phantoms of the Night," *National Geographic,* May 2001, 32–51; George Goodwin et al., *The Animal Kingdom* (New York: Greystone, 1954); C. A. W. Guggisberg, *Wild Cats of the World* (London: David and Charles, 1975); Rafael Hoogesteijn and E. Mondolfi, *The Jaguar* (Caracas: Ediciones Armitano, 1992); Diane Landau, ed., *Clan of the Wild Cats: A Celebration of Felines in Word and Image* (San Francisco: The Nature Company/Walking Stick Press, 1996); Les Line and Edward R. Ricciuti, *The Audubon Society Book of Wild Cats* (New York: Harry N. Abrams, 1985); R. A. Medellin, C. Equihua, C. L. B. Chetkiewicz, P. G. Crawshaw, A. Rabinowitz, K. H. Redford, J. G. Robinson, E. W. Sanderson, and A. B. Taber, *The Jaguar*

in the New Millennium (Mexico City: Universidad Nacional Autónoma de
México, Fonda de Cultura Económica, Wildlife Conservation Society, 2000);
S. Douglas Miller and Daniel D. Everett, eds., *Wild Cats of the World: Biology,
Conservation, and Management* (Washington, DC: National Wildlife Federa-
tion, 1986); Ronald M. Nowak, *Walker's Carnivores of the World* (Baltimore:
Johns Hopkins University Press, 2005); Richard Perry, *The World of the Jaguar*
(New York: Taplinger, 1970); David Quammen, *Monster of God: The Man-
Eating Predator in the Jungles of History and the Mind* (New York: Norton,
2003); Alan Rabinowitz, *Jaguar: One Man's Struggle to Establish the World's
First Jaguar Preserve* (New York: Arbor House, 1986); Paul Reddish, *The
Natural History and Ancient Civilizations of the Caribbean and Central Amer-
ica* (London: BBC Books, 1996); John Seidensticker and Susan Lumpkin,
Great Cats: Majestic Creatures of the Wild (Emmaus, PA: Rodale, 1991); K. L.
Seymour, "Panthera onca," *Mammalian Species,* 1989, 34; Mel Sunquist and
Fiona Sunquist, *Wild Cats of the World* (Chicago: University of Chicago Press,
2002); Elizabeth Marshall Thomas, *The Tribe of Tiger* (New York: Simon
and Schuster, 1994); W. C. Wozencraft in D. E. Wilson and D. M. Reeder, eds.,
Mammal Species of the World: A Taxonomic and Geographic Reference (Washington, DC: Smithsonian Institution Press in association with American Society of Mammalogists, 1993), 279–348.

I quote from the 1906 edition of the *New International Encyclopedia*
(New York: Dodd Mead, 94). The naturalist A. Starker Leopold is quoted
here and elsewhere from *Wildlife of Mexico* (Berkeley: University of California
Press, 1972). The 1830 description by J. R. Rengger is reported in Sunquist
and Sunquist, *Wild Cats of the World.* I relied greatly on Thomas, *The Tribe
of Tiger,* for descriptions of the behavior of large felids.

2. "IT PAYS US AGAIN AND AGAIN"

Descriptions of Belize, its history, and its conservation movement derive from
firsthand experience and interviews (1986–2008), as well as material reported
in my book with Kevin Shafer and Steele Wotkyns III, *Belize: A Natural Destination* (Santa Fe, NM: John Muir, 1991), and in Gary Hartshorn et al., *Belize:
Country Environmental Profile* (San José, Costa Rica: Trejos Hermanos, 1984).
They were enhanced by the work of Bruce Barcott, *The Last Flight of the Scarlet Macaw* (New York: Random House, 2008). The Aldous Huxley quotation
about British Honduras is found in *Beyond the Mexique Bay* (London: Chatto
and Windus, 1950, 35). I interviewed Belize Prime Minister George Price at his

Belize City home in November 1987 and later that month interviewed Therese Bowman in Dangriga.

Many of the Cockscomb facts and figures are from the article by A. R. Rabinowitz and B. G. Nottingham, "Ecology and Behavior of the Jaguar in Belize, Central America," *Journal of Zoology* 210 (1986): 149–159, as well as Rabinowitz's *Jaguar* and the 2003 National Geographic TV documentary *In Search of the Jaguar,* produced by Kate Churchill. I also quote from Lily Huang's *Newsweek* interview with Alan Rabinowitz, "Tiger Troubles," May 1, 2008, www.newsweek.com/id/135050. Other information came from April 2006–February 2008 personal communications with Rebecca Foster, Bart Harmsen, and Nicasio Coc (April 2006). I interviewed the Cockscomb caretaker Dan Taylor and the guide Ignacio Pop in November 1987.

The Victor González quotation is from Mahler, Shafer, and Wotkyns, *Belize.* The quotation from Charles Darwin is from *On the Origin of Species* (New York: Random House, 1999). Estimates on the number of jaguars in the wild are from Mel and Fiona Sunquist's *Wild Cats of the World.*

The conversation with Mark Pretti was in January 2001 at Ramsey Canyon. References to jaguars seen in Arizona historically are from *Borderland Jaguars: Tigres de la Frontera,* by David E. Brown and Carlos López González (Salt Lake City: University of Utah Press, 2001).

3. "AMONG ALL BIG CATS, WE KNOW LEAST ABOUT THEM"

Natural history information was gleaned from sources cited in Chapters 1 and 2.

Extinction of tiger subspecies is described in Quammen, *Monster of God,* and Ruth Padel, *Tigers in Red Weather: A Quest for the Last Wild Tigers* (New York: Walker, 2006). Eduardo Carrillo is quoted from "Tracking the Elusive Jaguar."

Unanswered questions about the jaguar and origin of its various names are discussed by Guggisberg, *Wild Cats of the World;* Julio Cesar Centeno, "The Cry of the Jaguar," *Environmental News Network,* 1997; and George Goodwin et al., *The Animal Kingdom* (New York: Greystone, 1954). The quotations from Louise H. Emmons are from Seidensticker and Lumpkin, *Great Cats.* The Natalie Angier comments are from her article "At Last Ready for Its Close-Up," *New York Times,* June 17, 2003. For information about general jaguar characteristics, I am indebted to Susan McGrath, "Top Cat," and Amy Linn, "Wild Cats Wild," *Audubon,* July–August 1993. Susan Morse is quoted on feline behavior by Kevin Hansen, *Bobcat: Master of Survival* (New York: Oxford University Press, 2007). Diane Landau comments on cat characteristics are from

Clan of the Wild Cats: A Celebration of Felines in Word and Image. I had personal communications with Kevin Hansen (August 2006–February 2009) about feline behavior and anatomy. Also contributing to this section were Murray E. Fowler and Zalmir S. Cubas, eds., *Biology, Medicine, and Surgery of South American Wild Animals* (Ames: Iowa State University Press, 2001), and Jack L. Childs, *Tracking the Felids of the Borderlands* (El Paso, TX: Printing Corner Press, 1998).

Warner Glenn's comments are from *Eyes of Fire;* Jeremy Kahn, "On the Prowl," *Smithsonian,* November 2007, 85–92; and personal communication. Much of the material about jaguars in the United States since 1996 is gleaned from personal interviews with Warner Glenn, Jack Childs, Emil McCain, and Sergio Ávila, as well as on-site research. I also benefited from Kahn's "On the Prowl" and newspaper articles by Sandra Blakeslee ("Gone for Decades, Jaguars Steal Back to the Southwest," *New York Times,* October 10, 2006) and Jeremy Voas ("Cat Fight on the Border," *High County News,* October 15, 2007, 9–15). I also attended meetings of the Jaguar Conservation Team and reviewed its reports in 2004, 2005, 2007, and 2008. The 1973 EPA language on endangered species is via www.epa.gov/regulations/laws/esa.html. Quotations from Tony Povilitis are from an Internet interview, "Tony Povilitis Q and A," Southwest Jaguars blog, February 2, 2008, www.swjags.wordpress.com/2008/02/02/tony-povilitis-qa/. Quotations and information from Alan Rabinowitz are from personal communication in 2004 and 2009, and articles by Paula MacKay ("Cockscomb: It's About Cats," *Wildlife Conservation,* November–December 2004, 30–34; and "Cockscomb in Context," *Wildlife Conservation,* November–December 2004, 35) and Rabinowitz ("Connecting the Dots: Saving the Jaguar Throughout Its Range," *Wildlife Conservation,* February 2006, 25–30).

4. "WE ALL FELT REALLY BLESSED"

The Jag Team and southern Arizona observations and conversations are from meetings attended in January and July 2004.

Alan Rabinowitz has discussed his childhood, youth, and stuttering in numerous interviews (see citations in earlier chapters). Particularly useful were a keynote speech to the Stuttering Foundation of America conference in Minneapolis (June 17, 2005, www.stutteringhelp.org/default.aspx?tabid=179&referrer=showText) and a National Public Radio interview (January 26, 2007, www.npr.org/templates/story/story.php?storyId=7043116).

Information about Maya Center came from conversations with Nicasio Coc, Aurora García, Emiliano Pop, and Ernesto Saqui in 2006. Jaguar field study information is from the WCS conservation Web site (www.savethejaguar .com). I interviewed Therese Bowman about Rabinowitz in 2005.

5. "WELL DRAWN AND UNMISTAKABLE"

Information about the Buenos Aires Wildlife Refuge and Baboquívari Mountains is from a January 2004 visit. The Dodd Mead *New International Encyclopedia* quotation is from the 1906 edition cited earlier. Natural history resources estimating jaguar ranges are those noted in Chapter 1 and 2, as well as Susan McGrath, "Top Cat," and the World Conservation Union, *IUCN Directory of Neotropical Protected Areas* (Dublin: Tycooly, 1983).

For details on jaguar sensory abilities, I drew particularly from K. L. Seymour, "Panthera onca," *Mammalian Species*, 1989, 34, and John L. Gittleman, ed., *Carnivore Behavior, Ecology, and Evolution* (Ithaca, NY: Comstock, 1989).

Descriptions of prehistoric jaguars and megafauna are found in Joseph L. Chartkoff and Kerry Kona Chartkoff, *The Archaeology of California* (Stanford: Stanford University Press, 1984); and Alan Rabinowitz, "Jaguars and Livestock: Living with the World's Third Largest Cat," in Woodroffe, Thirgood, and Rabinowitz, *People and Wildlife*, 278–285. The jaguar attack quotation is from Brown and López González, *Borderland Jaguars*, 128.

For background on jaguars and early indigenous cultures I benefited from personal communications with Steve Pavlik (2004–2005) and his article "Rohanas and Spotted Lions: The Historical and Cultural Occurrence of the Jaguar Among the Native Tribes of the American Southwest," *Wicazo Sa Review* 18 (Spring 2003): 157–175. Hopi aspects are described by Pavlik (personal communications, January 2004) and Brown and López González, *Borderland Jaguars*, as well as M. W. Billingsley, *Behind the Scenes in Hopi Land* (Tucson: University of Arizona Press, 1971). I had personal communications with Brad Draper (2004–2006) about jaguars in rock art and interviewed Cahuilla sources at the Agua Caliente Tribal Museum in Palm Springs, CA (2004).

Early U.S. jaguar sightings are described by Brown and López González, *Borderland Jaguars*, as well as by Peter Matthiessen, *Wildlife in America* (New York: Viking, 1959); Richard Perry, *The World of the Jaguar* (New York: Taplinger, 1970); Raymond J. Hock, "Southwestern Exotic Felids," *American Midland Naturalist* 53 (Spring 1955): 324–328; Vernon Bailey, *Mammals of New Mexico* (Washington, DC: U.S.D.A. Bureau of Biological Survey, 1931); E. W. Jameson Jr. and Hans J. Peeters, *California Mammals* (Berkeley: University of California Press, 1988); and Ernest Thompson Seton, *Lives of Game Animals,* vol. 2, pt. 1, *Cats, Wolves, and Foxes* (Garden City, NY: Doubleday, Doran,

1929). The quotation of Ignaz Pfefferkorn is from *Sonora: A Description of the Province* (Tucson: University of Arizona Press, 1989). I also cite Leopold, *Wildlife of Mexico.*

I relied on newspaper reports for accounts of jaguar attacks on zookeepers, including "'85 Attack at GR Zoo Led to Safety Review," *Grand Rapids Press,* December 26, 2007; "Jaguar Attack," *Fredericksburg News Post,* February 17, 2009; "Police Investigate Deadly Tiger Escape," *San Francisco Chronicle,* December 27, 2007. Statistics on mountain lion deaths are from Kevin Hansen, *Bobcat;* Bob Butz, *Beast of Never, Cat of God: The Search for the Eastern Puma* (Guilford, CT: Globe Pequot/Lyons, 2005); and David Baron, *The Beast in the Garden* (New York: Norton, 2004). I had e-mail correspondence with Peter G. Crawshaw Jr. (2004–2005). The Rabinowitz quotation is from "Jaguars and Livestock."

6. "THE MODEL FOR HOW TO LIVE"

I interviewed Guillermo Morales at Che Chem Ha in 1997. Denning habits of jaguars are described in Nowak, *Walker's Carnivores of the World.*

For background on jaguars and indigenous peoples, my primary sources were Michael Bright, *Man-Eaters* (New York: St. Martin's, 2002); Michael D. Coe, *Mexico* (New York: Thames and Hudson, 1982); Elizabeth P. Benson, ed., *The Cult of the Feline: A Conference in Pre-Columbian Iconography* (Washington, DC: Dumbarton Oaks, 1972); Mark Miller Graham, ed., *Reinterpreting Prehistory of Central America* (Niwot: University Press of Colorado, 1993); and Joseph Campbell, *The Power of Myth, with Bill Moyers,* ed. Joseph Flowers and B. Flowers (New York: Doubleday, 1988).

Quotations are from Quammen, *Monster of God,* and Perry, *The World of the Jaguar.* The article about the ancient Olmec with filed teeth is from Britt Peterson, "Just in Time for Halloween," *Discover,* October 2006, 12.

7. "JAGUARS POSSESS THE POWER OF GOD"

In addition to personal research and interviews conducted in Mexico and Central America (1992–2008), I drew particularly from Michael D. Coe, *Final Report: An Archaeologist Excavates His Past* (New York: Thames and Hudson, 2006); Pat Culbert, "Maya Curiosities," *Archaeology,* November 2002, 94; Daniel D. Brinton, *The Maya Chronicles* (Philadelphia: Brinton, 1882); J. Eric S. Thompson, *Maya History and Religion* (Norman: University of Oklahoma Press, 1970); Mary Miller and Karl Taube, *An Illustrated Dictionary of the Gods* and Symbols of Ancient Mexico and the Maya (London: Thames and Hudson, 1993); Ann and Myron Sutton, Among the Maya Ruins (New York: Rand McNally, 1967); Paul Reddish, The Natural History and Ancient Civilizations of the Caribbean and Central America (London: BBC Books, 1996); John Noble Wilford, "On Ancient Walls, a New Maya Epoch," New York Times, May 16, 2006; Linda Schele, David Freidel, and Joy Parker, Maya Cosmos: Three Thousand Years on the Shaman's Path (New York: Quill/Morrow, 1993); Linda Schele and David Freidel, A Forest of Kings: The Untold Story of the Ancient Maya (New York: Quill/Morrow, 1990); Gene S. Stuart and George E. Stuart, Lost Kingdoms of the Maya (Washington, DC: National Geographic Society, 1993); John Lloyd Stephens, Incidents of Travel in Central America, Chiapas, and Yucatán (Washington, DC: Smithsonian Institution Press, 1993); Norman Hammond and Gordon R. Willey, eds., Maya Archaeology and Ethnohistory (Austin: University of Texas, 1979); and Charles Gallenkamp, Maya: The Riddle and Rediscovery of a Lost Civilization (New York: Penguin USA, 1987). I also quote from Alan Rabinowitz's Jaguar.

For information about the Aztec I used works by George C. Vaillant (The Aztecs of Mexico [Garden City, NY: Doubleday, 1950]), Bernal Díaz del Castillo (The Discovery and Conquest of Mexico, 1517–1521 [New York: Farrar, Straus, and Cudahy, 1956]), and Muriel Porter Weaver (The Aztecs, Maya, and Their Predecessors [New York: Academic Press, 1981]). Information about the Chavín temple is from an article entitled "Researcher Discovers Old Jaguar Relief," Yomiuri Shimbun, January 12, 2006.

8. "BLOOD OF THE VALIANT"

I quote here from Nowak, Walker's Carnivores of the World, and Rabinowitz, People and Wildlife. Information sources about pelts, Chinese medicine, and wildlife trade include Sharon Begley, "Extinction Trade," Newsweek, March 10, 2008, 47–49; Barun Mitra, "Sell the Tiger to Save It," New York Times, August 15, 2006; Richard Ellis, Tiger Bone and Rhino Horn: The Destruction of Wildlife for Traditional Chinese Medicine (Washington, DC: Island, 2005); Keith Schneider, "Mediating the Federal War on Wildlife," New York Times, June 9, 1991; N. Smith, "Spotted Cats and the Amazon Skin Trade," Oryx 13, no. 4 (1976): 362–371; and Peter Matthiessen, Tigers in the Snow (New York: North Point /Farrar, Straus and Giroux, 2000). Background on CITES is from Barcott, The Last Flight of the Scarlet Macaw. I quote F. Bruce Lamb, Wizard of the Upper Amazon (Berkeley: North Atlantic, 1971).

My sources on jaguar hunting include personal interviews cited earlier (2004–2008), including those with David Brown, Jack Childs, Dave Stourzdas, Emil McCain, Warner Glenn, and Michael Robinson. I also relied on documents provided by the U.S. Fish and Wildlife Service and an Internet interview with Ron Thompson (September 2007, "Ron Thompson Q & A," Southwest Jaguars blog, September 18, 2007, www.swjags.wordpress.com/2007/09/18/ron -thompson-qa/). Klump case news reports included Tim Vanderpool, "Klump Clan," *San Pedro Valley Sun-News,* May 26, 2004, and "Two Guilty of Selling Hides," *New York Times,* November 27, 1998. Information on jaguar bounties and early hunting of big cats draws from Brown and López González, *Border-land Jaguars;* J. P. S. Brown, *The Forests of the Night* (Lincoln, NE: iUniverse, 2008); Neil B. Carmony, *Onza: The Hunt for a Legendary Cat* (Silver City, NM: High-Lonesome, 1995); Robert McCurdy, *Life of the Greatest Guide* (Phoenix: Blue River Graphics, 1981); and North American Big Game Awards, *North American Big Game: Seventh Edition* (Chicago: Donnelly, 1977). I quote from Theodore Roosevelt, "A Jaguar-Hunt on the Taquary," in *Through the Brazil-ian Wilderness* (New York: Scribner's, 1926).

I quote George Schaller from his memoir, *A Naturalist and Other Beasts: Tales from a Life in the Field* (San Francisco: Sierra Club Books, 2007); Sandra Cavalcanti from a TV documentary produced by Kate Churchill, *In Search of the Jaguar* (Washington, DC: National Geographic, 2003); Eric Gese from a Jaguar Conservation Team talk delivered in May 2007 in Douglas, AZ; and Susan McGrath from her article "Brazil's Wild Wet, the Pantanal Wetland," *National Geographic,* August 2005, 52–69.

American University's James R. Lee posts data about pelt smuggling at www.american.edu/TED/class/allcrime.htm. The Conservation International citation is from press information at www.conservationinternational.org/ discover/science/discovery/pantanal/dispatches/Pages/dispatches.aspx. Background on Sasha Siemel is from his memoir, *Tigrero* (New York: Prentice-Hall, 1953), and his memorial Web site, www.sashasiemel.com.

In describing the Pantanal, I had e-mail correspondence with Peter Craw-shaw Jr. (February 2004) and Don North (April 2006). I also relied on Douglas H. Chadwick's "Phantoms of the Night," *National Geographic,* May 2001, 32–51.

9. "HE BELIEVES HE *IS* A JAGUAR"

For information about jaguars, tribal cultures, shape-shifting, and shamanism I am particularly indebted to Liz Rymland (2004 personal interview) and the following: Thompson, *Maya History and Religion;* Gerardo Reichel-Dolmatoff,

Amazonian Cosmos: The Sexual and Religious Symbolism and the Tukano Indians (Chicago: University of Chicago Press, 1971); Gerardo Reichel-Dolmatoff, *The Shaman and the Jaguar* (Philadelphia: Temple University Press, 1975); and Tom Huth, "The Highs of Ecuador," *Condé Nast Traveler,* September 1993, 103–177.

The segment on the Usumacinta and Lacandón is based on personal experiences and interviews in the region (1994, 1996), as well as writings by Christopher Shaw (*Sacred Monkey River: A Canoe Trip with the Gods* [New York: Norton, 2000]), Maud Worcester Makenson (*The Book of the Jaguar Priest* [New York: Henry Schuman, 1951]), Jon Christopher Crocker (*Vital Souls: Bororo Cosmology, Natural Symbolism and Shamanism* [Tucson: University of Arizona Press, 1985]), and Victor Perera and R. Bruce (*The Last Lords of Palenque: The Lacandón Mayas of the Mexican Rain Forest* [Berkeley: University of California Press, 1982]). Also useful was R. Cooke's "Animal Icons and Pre-Columbian Society: The Felidae, with Special Reference to Panama," in Graham, *Reinterpreting Prehistory of Central America,* 169–208.

I quote Robert Carneiro, *Muse of History and Science of Culture* (New York: Springer, 2000), about shamans, as well as Theodor Koch-Grunberg, "Keeping It Oral: A Yekuana Ethnology," *American Ethnologist* 13, no. 3 (1986): 413–429. The ayahuasca description is from Manuel Villavicencio, *Geografía de la República del Ecuador* (New York: Robert Craighead, 1858). I quote from an article by Diane Hamilton ("Tobacco Shamans," *Los Angeles Times,* October 26, 1986), as well as from works by Johannes Wilbert (*Tobacco and Shamanism in South America* [New Haven: Yale University Press, 1993]), Graham (*Reinterpreting Prehistory of Central America*), Martín Prechtel (*Secrets of the Talking Jaguar* [New York: Tarcher/Putnam, 1998]), and Gyles Iannone ("Howl from the Bowel," *Discover,* October 1995, 11). Background on the Matsés tribe is from Benedict Allen, *Through Jaguar Eyes: Crossing the Amazon Basin* (London: Flamingo, 1995).

10. "THERE IT IS; I'M GOING TO SHOOT IT"

Belize conservation information is from various published sources previously cited, personal interviews (mainly in April 2006 with Celso Poót, Sharon Matola, Vladimir Rodríguez, Ryan Phillips, Chris Hatten, and Carolyn and Bruce Miller), and reports issued by the Belize Department of Forestry. Useful information on Belize wildlife came from Carolyn M. Miller and Bruce W. Miller, *Exploring the Rainforest* (Chan Chich, Belize: Chan Chich Lodge,

1988). I quote Roger B. Caras, *A Cat Is Watching* (New York: Fireside, 1990), and George Schaller as reported in Padel, *Tigers in Red Weather.*

11. "COWS ARE MORE IMPORTANT THAN CATS"

The opening story and quotations are from David E. Brown, "Revival for el Tigre?" *Defenders* 66, no. 1 (2000): 27–35. I had personal communications with Roberto Águilar in May–June 2007. Statistics on jaguars in Sonora are from interviews and personal communication with David E. Brown (May 2004 and February 2007), Emil McCain (July–August 2004, February 2006, May–June 2007), and Peter Warshall (November 2007 and January 2009).

I quote Sandra Cavalcanti from Churchill, *In Search of the Jaguar,* and Parisina Malatesta, "Shifting Sands and Colonial Continuity," *Américas,* January 1996, plus a passage from P. H. Fawcett in *Lost Trails, Lost Cities: An Explorer's Narrative by Colonel P. H. Fawcett,* ed. Brian Fawcett (New York: Funk and Wagnalls, 1953). Specifics about Venezuela's jaguars are from Rafael Hoogesteijn and E. Mondolfi, "Notes on the Biology and Status of the Jaguar in Venezuela," in *Cats of the World: Biology, Conservation, and Management,* ed. S. D. Miller and D. D. Everett (Washington, DC: National Wildlife Federation, 1986), and Stan Steiner, *In Search of the Jaguar: Growth and Paradox in Venezuela* (New York: Times Books, 1979). The 2007 *carne de onca* account is from Henri Paget, news.ninemsn.com.au/article.aspx?id=181963. Quotations attributed to Brown and López González are from *Borderlands Jaguars,* as are details about jaguars preying on cattle. Quotations from Juan Carlos Bravo are from "Juan Carlos Bravo Q&A," Southwest Jaguars blog, November 7, 2007, swjags .wordpress.com/2007/11/07/juan-carlos-bravo-qa/. My primary interview with Sergio Ávila and Emil McCain was in July 2004. I interviewed Octavio Rosas-Rosas in person and via e-mail, 2004–2008. Accounts of the Sierra Madre Occidental are from a March 2005 visit.

12. "WE JUST STOPPED SEEING THEM"

Quotations from McCain and Ávila are primarily from personal interviews cited earlier and e-mail correspondence July–December 2004. Sky Island Alliance data on Sonoran jaguars was provided by Sergio Ávila. David E. Brown was interviewed at Arizona State University (March 2, 2004). Information and quotations relating to the Northern Jaguar Reserve are from Diane Hadley (personal interview and e-mail correspondence, October–November 2007), Warshall (November 2007 and January 2009), and Brian Miller (e-mail correspondence and interview, May–June 2007). I had personal communications with T. Luke George (August 2004) and Bill Van Pelt (January and July 2004). Policies on jaguar reintroduction and reports on their conservation were released at Jaguar Conservation Team meetings by the Arizona and New Mexico game departments in 2006. Other material on the border fence is from Emil B. McCain and Jack L. Childs, "Evidence of Resident Jaguar in the Southwestern United States and the Implications for Conservation," *Journal of Mammalogy* 89, no. 1 (2008): 1–10; Lee Morgan II, *The Reaper's Line: Life and Death on the Mexican Border* (Tucson, AZ: Rio Nuevo, 2006); Jack L. Childs and Anna Mary Childs, *Ambushed on the Jaguar Trail: Hidden Cameras on the Mexican Border* (Tucson, AZ: Rio Nuevo, 2008); Jack Childs and Emil McCain, various reports of the Borderlands Jaguar Detection Project, 2005–2008; and Tom Dollar, "El Tigre," *Wildlife Conservation,* November–December 2004, 23–29. Quotations and background on the fence are also from personal communication with Peter Warshall (November 2007 and January 2009), Jack Childs (January and July 2004, May 2007, July 2008), Emil McCain (July–August 2004, February–March 2006, August–September 2007), Brian Miller (May–June 2007), Warner Glenn (May–June 2007, August–September 2007), Diane Hadley (October–November 2007), and Sergio Ávila (July 2004, May 2007, July 2008). Information about Macho B's capture is from a February 19, 2009, news release by Arizona's Department of Game and Fish.

13. "TO ENSURE OUR NAMESAKE IS PROTECTED"

Sources for Jaguar automobile history were primarily corporate documents, spokeswoman Rosemary Mariniello, and archivist Mike Cook (e-mail correspondence with both). Other sources include the Wildlife Conservation Society (www.savethejaguar.com); Jaguar Conservation Trust (www.keycast.com/keycast/jaguar/jct/); "Tata Buys Jaguar in £1.15bn Deal," March 26, 2008, news .bbc.co.uk/1/hi/business/7313380.stm; Evan McMullen, "Jaguar Name Grew out of Prewar Mistake: The SS Model," *Seattle Post-Intelligencer,* September 24, 2004, seattlepi.nwsource.com/wheels/191990_jaguar24.html; Alan Rabinowitz ("Jaguar Cars Is Helping Save the Largest Cat in the Americas," special advertising section sponsored by Jaguar Cars, *Wildlife Conservation,* May–June 2005), "Jaguar Turns Winslet's Body into Sleek New Car," November 20, 2006, elina.actressarchives.com/news.php?id=2909; Kim Murphy and Henry Chu, "A Classy Act Relocates: The Jaguar Brand's Passage to India Is Steeped in Irony," *Los Angeles Times,* March 22, 2008.

Text on human preoccupation with cats and sports team mascots was
informed by Diana Landau, *Clan of the Wild Cats;* Dodd-Mead's *New International Encyclopedia;* www.en.wikipedia.com/jaguar, under the heading
"Contemporary culture"; Stephanie Chen, "Animal House: College Mascots
Get Luxury Digs," *Wall Street Journal,* August 15, 2007, online.wsj.com/
article/SB118713865671897991.html; and various personal interviews.

14. "SIGA EL PISTO"

Information relating to jaguars in Guatemala was gathered from personal research and interviews there (1992–2006). Primary sources included Roan Balas
McNab, Anthony J. Novack, Gabriela de la Hoz, and Hector Trujillo. I also
benefited from Roan Balas McNab, "A Preliminary Assessment of Jaguar Distri-
butions in Guatemala," unpublished internal report, Wildlife Conservation
Society, February 28, 1999; Anthony J. Novack, "Impacts of Subsistence Hunting upon Jaguar and Puma," M.S. thesis, University of Florida, 2003; Manuel
Roig-Franzia, "Linked Killings Undercut Trust in Guatemala," *Washington
Post,* March 23, 2007, www.washingtonpost.com/wp-dyn/content/article/2007/
03/22/AR2007032201903.html; and "Conservation: Jaguar Research Projects,"
Wildlife Conservation Society news release, 2006, www.savethejaguar.com/
jag-index/jag-conservation/jag-research?preview=&psid=&ph=class%25.

15. "LIVING IN THE SAME PLACE IT ALWAYS HAS"

For information about Cockscomb, I used multiple sources cited in earlier
chapters, as well as interviews conducted via e-mail and in or near the sanctuary
(notably with Bart Harmsen, Rebecca Foster, Emiliano Pop, Ferdie Yau, Nicasio Coc, Ernesto Saqui, and Aurora García). I found some background and a
quotation from Saqui in Donald G. Schueler, *The Temple of the Jaguar: A Personal Journey of Hope and Renewal* (San Francisco: Sierra Club, 1993).

16. "PRETTY WELL HUNTED OUT"

Ocelot natural history was described by Rebecca Foster in interviews and by
Laura Tangley, "Cat on the Spot," *National Wildlife,* April–May 2006, 31–37.
Interviews with Rebecca Foster, Emiliano Pop, Nicasio Coc, Ferdie Yau, and
Julian Clayton were conducted in Belize during April 2006. Bart Harmsen was
interviewed via e-mail in 2006–2007. Emory King was interviewed in Belize
during May 2000.

17. "THESE ANIMALS COULD BECOME WONDERFUL TEACHERS"

The principal interview with Sharon Matola was done at the Belize Zoo in April 2006. Other interviews were in person or via e-mail in 2006–2008 and during the 1990s. A helpful description of Matola was written by Bruce Barcott in *The Last Flight of the Scarlet Macaw*. Useful background on the zoo came from Barcott as well as from interviews with Richard and Carol Foster, the facility's neighbors, and Celso Poót at the Tropical Education Center.

The American Zoo and Aquarium Association (Jaguar Species Survival Plan, www.jaguarssp.org) provided information about jaguars and other large felids in zoos, as did the Phoenix Zoo's Roberto Águilar (personal communications, May–June 2007). I interviewed Barbara Dicely of Leopards Etc. in California in November 2006. The breeding of zoo animals is discussed by David Quammen in *Wild Thoughts from Wild Places* (New York: Touchstone, 1998).

18. "IT'S GOOD IF IT'S DEAD"

Following my visit to Costa Rica, during April 2006, supplemental interviews were with Emil McCain, Eduardo Carrillo, and Ryan Phillips, as cited earlier. I benefited from articles about Costa Rica published in the Latin America edition of the *New York Times* (Hillary Rosner, "Monos mueren por exceso de lluvia en Costa Rica," March 19, 2006), *National Geographic* ("Jaguars: Phantoms of the Night," May 2001), *Natural History* ("Tracking the Elusive Jaguar," May 2007, and John Roach ("Elusive Jaguars Remain a Mystery," November 30, 2003, www.newsnationalgeographic.com/news/2003/11/1125_031125_jaguars .html), and *Wildlife Conservation* ("Connecting the Dots," February 2006), as well as Joanne Omang ("In the Tropics, Still Rolling Back the Rain Forest Primeval," *Smithsonian,* April 2004, 56–67) and reporting on the Wildlife Conservation Society Web site ("Costa Rica Programs," www.wcs.org/ globalconservation/latinamerica/mesoamerica/costarica, and "Jaguars in Costa Rica," www.wcs.org/353624/199500). The official Web site for Corcovado National Park (www.costarica-nationalparks.com/corcovadonationalpark.html) provided background, as did literature supplied by various conservation groups working on the Osa Peninsula.

19. "A FLAGSHIP SPECIES FOR CONSERVATION"

Quotations and background related to jaguar conservation are from Alan Rabinowitz ("An Asian Pygmy and His Pal from Far Rockaway," National Public

Radio, January 26, 2007, www.npr.org/templates/story/story.php?storyid=
7043116; and "National Geographic Weekend," *National Geographic News,*
December 22, 1007, www.nationalgeographic.com/radio/episodes/episode-117
.html); Peter Warshall (November 2007 interview); Roan McNab (April 2006
interview), George Schaller (*A Naturalist and Other Beasts*), Emil McCain
(various interviews and e-mail correspondence, July 2004–September 2007, as
well as unpublished reports of the Borderlands Jaguar Detection Project, June
2006 and January 2007), Jack Childs (with Emil McCain, *Journal of Mammalogy,* and, with Anna Childs, *Ambushed on the Jaguar Trail*), and the Wildlife
Conservation Society ("Central America Agrees to Jaguar Corridor," news release, May 23, 2006).

Information about highway crossings for animals and roadkill statistics
came in part from Peter Aleshire, "Safe Crossing," *High Country News,* November 12, 2007; and the Wildlands Project Web site (www.twp.org/cms/page1131
.cfm).

In writing about Mexican jaguars I benefited from the Friends of Calakmul Web site (www.rainforest2reef.org); e-mail correspondence with Gerardo
Ceballos (February 2007); Lane Simonian, *Defending the Land of the Jaguar:
A History of Conservation in Mexico* (Austin: University of Texas Press, 1995);
Barbara Kingsolver, "The Way to Nueva Vida," *Sierra,* September–October
2003, 34–54; Eric Niiler, "Into the Jaguar's Den," *Scientific American,* September 2001, 20–22; and Charles Bergman, "The Cat That Walks by Itself," *Smithsonian,* December 2004, 55–62).

My writings about jaguars in Brazil were informed by "Looking at Lula,"
E, September–October 2005; Jennifer Shatwell, "Ranchers Give Refuge to
Pantanal Jaguars," Conservation International, www.conservation.org/FMG/
Articles/Pages/ranchers_refuge_jaguars.aspx; Peter G. Crawshaw and H. B.
Quigley, "Jaguar Spacing, Activity, and Habitat Use in a Seasonally Flooded
Environment in Brazil," *Journal of Zoology* 223 (1991): 357–370; H. B. Quigley
and Peter G. Crawshaw, "A Conservation Plan for the Jaguar in the Pantanal,"
Biology and Conservation 61 (1992): 149–157; Larry Rohter, "Brazil Gambles
on Monitoring of Amazon Loggers," *New York Times,* January 14, 2007; George
B. Schaller, "Mammals and Their Biomass on a Brazilian Ranch," *Arquivos
de Zoologia* 31 (1983): 1–36; Hoogesteijn and Mondolfi, *The Jaguars;* and George
B. Schaller and Peter G. Crawshaw, "Movement Patterns of Jaguar," *Biotropica*
12 (1980): 161–168. Several quotations from Sandra Cavalcanti and Marcos Moraes and a reference to Fernando Azevedo derive from J. Madeleine Nash, "Can
They Stay Out of Harm's Way?" *New York Times,* January 1, 2008.

Reporting related to Paraguay was aided by personal interviews with Anthony Novack and Don North (2006), as well as by Novack's "Impacts of Subsistence Hunting."

Sources of information about new jaguar research include Eric Gese's Jaguar Conservation Team talk; Peter Aldous, "Big Cats Hit by Photographers' Cheap Trick," *New Scientist,* February 18, 2006; and "Mother Lode of Jaguars Discovered in Bolivia Park," Wildlife Conservation Society news release, May 17, 2004.

20. "THE MOTHER LIQUOR FROM WHICH WE HAVE COME"

Panama Canal jaguar swims are discussed in two articles: "Scientists Turn to a Satellite to Save Guatemala's Jaguars," *New York Times* (Reuters), May 22, 2005; and Terry McCarthy, "Nowhere to Roam," *Time,* August 23, 2004, 44–53. My visit to Panama was in April–May 2006, and some information came from informal interviews with guides, park wardens, missionaries, and residents.

The 1922 *National Geographic* quotation from David Fairchild and some background on Panama's natural resources are derived from Elizabeth Royte, *The Tapir's Morning Bath: Mysteries of the Tropical Rain Forest and the Scientists Who Are Trying to Solve Them* (Boston: Houghton Mifflin, 2001). I drew material about conservation of big cats from Richard Ives, *Of Tigers and Men* (New York: Avon, 1996); and Corey J. Meacham, *How the Tiger Lost Its Stripes* (Orlando: Harcourt Brace, 1997).

This chapter was informed by two TV documentaries about jaguars: Churchill, *In Search of the Jaguar;* and Richard Foster and Carol Farneti Foster, producers, in association with WNET-TV New York, "Jaguar: Year of the Cat," *Nature* (PBS-TV, 1995).

Peter Matthiessen is quoted here from *The Snow Leopard* (New York: Penguin, 1978), his account of a Himalayan quest he undertook with George Schaller.

21. "TO SEE ONE AT ALL IS A LIFETIME EXPERIENCE"

Some of the Barro Colorado description and quotations are from Royte, *The Tapir's Morning Bath.* Global warming and tropic ecology are described by Ed Ayres, *God's Last Offer* (New York: Four Walls Eight Windows, 1999); and Jeffrey P. Cohn, "Call of the North," *Américas,* September–October 1988, 39–44.

Media reports consulted included those written by Hillary Rosner ("Monos mueren por exceso de lluvia en Costa Rica," *New York Times,* Latin

America ed., March 19, 2006) and Jon Hamilton ("Rainfall Shortages Threaten Costa Rica Power," *Morning Edition,* National Public Radio, February 11, 2008, www.npr.org/templates/story/story.php?storyid=18832252).

An article by Mike Murphy provided information about the Formosan clouded leopard and related Rabinowitz research ("Strategies for Protecting Frail Ecosystem," *Taipei Times,* April 15, 2005).

Reports on big cat sightings in unusual places were gleaned from numerous sources, including Ron Toft, "Big Cats: The Truth Is Out There," *London Financial Times,* March 13, 2004; Big Cat Society (www.britishbigcats.org); "Big Cat May Be Stalking Western Sydney," the Australian Broadcasting Corporation, www.abc.net.au/news/stories/2006/07/27/1698320.htm; and "Panther Reported in Barfield," WGNS Radio, December 18, 2007, wgns.wordpress .com/2007/12/18/jaguar-reported-in-barfield.

Particularly useful in describing predator conservation were Justina C. Ray, Kent H. Redford, Robert S. Steneck, and Joel Berger, *Large Carnivores and the Conservation of Biodiversity* (Washington, DC: Island Press, 2005); and Quammen, *Monster of God* and *Wild Thoughts from Wild Places. Wild Thoughts* also is the source of the Colombian jaguar myth. World Wildlife Federation jaguar population estimates are from www.worldwildlife.org/what/howwedoit/ policy/WWFBinaryitem8746.pdf.

Saving and Studying Jaguars

Protecting and investigating jaguars and their diverse habitats involves an enormous, expensive, and complicated set of tasks. Many organizations, agencies, and individuals are committed to such endeavors. For some the effort is one of many activities in which they are immersed. Among these groups are the Wildlife Conservation Society and Panthera Foundation nonprofit organizations that sponsor important research on behalf of jaguars. WCS and Panthera were not involved directly in the production or writing of this book, but I am contributing a percentage of my royalties to their jaguar conservation programs.

FOR ADDITIONAL INFORMATION ABOUT JAGUARS

Big Cat Rescue
12802 Easy Street
Tampa, FL 33625
813-920-4130
www.bigcatrescue.org or www.bigcats.com

Cat Survival Trust for Endangered Species
The Centre, Codicote Road
Welwyn, Hertsfordshire AL6 9TU
England
44-0-1438-716873
www.catsurvivaltrust.org

Center for Biological Diversity
P.O. Box 710
Tucson, AZ 85702
602-628-9909
www.biologicaldiversity.org

In Search of the Jaguar, one-hour National Geographic TV special, November 2003. Produced and directed by Kate Churchill. Nama Productions, www.namaproductions.com/pages/projects.html

"Jaguar: Year of the Cat," one-hour PBS-TV *Nature* segment, 1995. Produced
by Richard Foster and Carol Farneti Foster in association with WNET-TV,
New York.
www.shopthirteen.org/product/show/29656

Southwest Jaguars Blog (Bill Rejebian)
www.swjags.wordpress.com

World Conservation Union/IUCN Red List
Rue Mauverney 28
Gland 1196
Switzerland
41-22-999-0000

PLACES YOU MIGHT SEE A JAGUAR

Belize Lodge and Excursions
Toledo District, Belize
888-292-2462 or 501-223-6324
www.belizelodge.com

Belize Zoo
P.O. Box 1787
Belize City, Belize
501-220-8004
www.belizezoo.org

Chan Chich Lodge
P.O. Box 37
Belize City, Belize
800-343-8009 or 501-223–4419
www.chanchich.com

Cockscomb Basin Wildlife Sanctuary
Belize Audubon Society
Maya Center, Belize
www.belizeaudubon.org/protected_areas/cockscomb-basin-wildlife-sanctuary
.html

Corcovado National Park
Osa Peninsula, Costa Rica
www.costarica-nationalparks.com/corcovadonationalpark.html

(Calakmul Biosphere, Yucatán, Mexico)
Friends of Calakmul/Rainforest2Reef
P.O. Box 735
Tahoe City, CA 96145
650-430-4089
www.rainforest2reef.org

(Jaguars in U.S. zoos)
American Zoo and Aquarium Association
www.jaguarssp.org

Programme for Belize/Rio Bravo Conservation and Management Area
P.O. Box 749
Belize City, Belize
501-227-5616
www.pfbelize.org/facilities.html

Tikal National Park
El Petén, Guatemala
www.tikalpark.com

SOME ORGANIZATIONS AND AGENCIES INVOLVED IN
SAVING JAGUARS

Arizona Game and Fish Department, Jaguar Conservation Project
5000 W. Carefree Highway
Phoenix, AZ 85086
602-942-3000
www.azgfd.gov/w_c/es/jaguar_management.shtml

Borderlands Jaguar Detection Project
1165 W. Hawk Way
Amado, AZ 85645
www.borderjag.org

Conservation International
2011 Crystal Drive, #500
Arlington, VA 22202
703-341-2400
www.conservation.org

Defenders of Wildlife
1130 17th Street, NW
Washington, DC 20036
800-385-9712
www.defenders.org

Jaguar Conservation Trust
http://www.keycast.com/keycast/jaguar/jct/Default.asp?flashversion=9&

Naturalia (Sonora office)
Horquillas 43
Colonia, Santa Fe
Hermosillo, 83249
Sonora, México
55-55-59-56-96
www.naturalia.org.mx (Spanish only)

Northern Jaguar Project
2114 W. Grant Road, #121
Tucson, AZ 85745
520-623-9653, ext. 5
www.northernjaguarproject.org

Panthera Foundation
8 West 40th St.
18th Floor
New York, NY 10018
646-786-0400
www.panthera.org

Pro-Natura
Aspérgulas 22 (Antes Pino)
Colonia San Clemente CP 01740
Mexico City, Mexico
52-55-56-35-50-54, ext. 107
www.pronatura.org.mx

Red Yaguareté Argentina
www.jaguares.com.ar (Spanish only)

Sky Island Alliance
P.O. Box 41165
Tucson, AZ 85717
520-624-7080
www.skyislandalliance.org

Wildlands Project, Southwest Field Office
P.O. Box 16213
Portal, AZ 85632
575-557-0155
www.twp.org

Wildlife Conservation Society
2300 Southern Blvd.
Bronx, NY 10460
718-220-5100
www.wcs.org or www.savethejaguar.com

Several of these groups allow donors to "adopt" jaguars through charitable contributions. I encourage you to do so.

Acknowledgments

Hundreds of individuals and entities contributed to the creation of this book. Space restricts me to the mention of only a few of them, noted here in alphabetical order:

Roberto Águilar, Amerind Foundation, Stacey Austin, Sergio Ávila, Eileen Bailey, David Baron, Jean Thomson Black, Sandy Blakeslee, Marjorie Boltz, Therese Bowman, Laurie Brooks, David E. Brown, Stephen H. Buck, Rita Cadena, John and Carolyn Carr, Eduardo Carrillo, Gerardo Ceballos, Larry Cheek, Jack and Anna Mary Childs, Nicasio Coc, Peter Crawshaw Jr., Scott Davis, William deBuys, Brad Draper, Jay Dusard, Carol Farneti Foster, Rebecca Foster, Richard Foster, Aurora García, Lois Gilbert, Warner and Wendy Glenn, Diana Hadley, Kevin Hansen, Barbara Hanson, Chris Hatten, Dan Heaton, Janet Hughes, Susan Ives, Maritza Juárez, Emory King, Nasario Ku, Nicky Leach, Carlos López González, Charlie Luthin, Jane Susan MacCarter, Ron Mader, Don Mahler, Sharon Matola, Emil McCain, Ruth McGuirk, Roan McNab, Carolyn and Bruce Miller, Patty Moosbrugger, John Morehead, Penelope Mudd, Rose Najia, Lisa Noble, Tom Noble, Don North, Anthony Novack, Mark and Nico at Ojo del Mar, Nina Orda, Daniel Owsiany, Steve Pavlik, Victor Perera, Ryan Phillips, Celso Poót, Emiliano Pop, Mark Pretti, Sebastian Rabes, Alan Rabinowitz, Michael Robinson, Vladimir Rodríguez, Octavio Rosas-Rosas, Ernesto Saqui, Martha Schumann, Mark Setterland, Chris Shaw, Dave Strozdas, Ron Thompson, Don Usner, Abelino Valle, Vallecitos Mountain Refuge, Bill Van Pelt, John Ware, Peter Warshall, Gordon Whiting, Wildlife Conservation Society, and Steele Wotkyns III.

I offer a deep bow of appreciation to all. The deepest bows go to my partner in life and love, Stacey Austin, and our home's *Felidae* representative, Chumley the cat.

Index